6th Asian Conference on Environmental, Industrial and Energy Engineering (EI2E 2022)

IOP Conference Series: Earth and Environmental Science
Volume 1044

Online
20 May 2022

ISBN: 978-1-7138-5822-5
ISSN: 1755-1307

Printed from e-media with permission by:

Curran Associates, Inc.
57 Morehouse Lane
Red Hook, NY 12571

Some format issues inherent in the e-media version may also appear in this print version.

Institute of Physics
Dirac House, Temple Back
Bristol BS1 6BE UK

Phone: 44 1 17 929 7481
Fax: 44 1 17 920 0979

techtracking@iop.org

Additional copies of this publication are available from:

Curran Associates, Inc.
57 Morehouse Lane
Red Hook, NY 12571 USA
Phone: 845-758-0400
Fax: 845-758-2633
Email: curran@proceedings.com
Web: www.proceedings.com

TABLE OF CONTENTS

IOP Conf. Series: Earth and Environmental Science 1044 (2022) 011001 doi:10.1088/1755-1315/1044/1/011001

PREFACE

The 6th Asian Conference on Environmental, Industrial and Energy Engineering (EI2E 2022) in conjunction with the 2022 6th International Conference on Environmental and Energy Engineering (IC3E 2022) were co-organized by the Asia Pacific Institute of Science and Engineering (APISE) and the International Society for Environmental Information Sciences (ISEIS). Due to recent pandemic, it was finally held as an online conference via Tecent Meeting Software on May 20, 2022. The decision to hold the virtual conference was made in compliance with many restrictions that were imposed by countries around the world. These restrictions were made to minimize the risk of spreading the COVID-19. Some people suggest to postpone the conference, however, we prefer to use online meeting because of reasons:(1) Most of the authors want to publish their articles as scheduled, as some of them need it before graduation;(2) We don't know when it will be good enough to hold the conference as usual. The online conference was a great success. The participants were from a number of countries (Canada, India, Chile, Saudi Arabia, Germany, Uzbekistan, China, etc.) with diverse backgrounds, forming a coherent environment for facilitating tremendous learning, sharing and collaborating opportunities

Energy and environment are closely interrelated. Energy development is associated with many environmental concerns, where effective approaches for emission minimization are desired. At the same time, a number of innovative technologies for mitigating pollutant/carbon emissions and managing the related risks are being developed. EI2E 2022 is targeted on providing opportunities to bring together the related researchers to share their most recent research achievements in these fields, and thus promote more advanced research.

On the behalf of the conference organizing committee, I would like to express my great appreciation to the three keynote speakers (40 minutes each, including Q&A). In addition, the conference included two oral and one poster sessions. In the oral sessions, each presentation was allotted 15 minutes. At the end of each session, the participants were engaged in discussions for future collaborations. A group photo was taken at the conference.

Through peer reviews undertaken by conference committee members and international experts, 14 outstanding papers out of 32 papers are identified to be included in this book. As conference proceedings, this book contains two chapters, including (1) Advances in Energy and Environmental Engineering, and (2) Modeling of Resource and Environmental Systems. provides selected papers as presented at the conference.

I hope that readers of the proceedings will find that the presented papers are valuable references for supporting their research and development in the fields of energy and environmental research. Thank you very much, and we look forward to seeing you at EI2E 2023.

Gordon Huang, University of Regina, Canada
Conference Chair

IOP Conf. Series: Earth and Environmental Science 1044 (2022) 011001 doi:10.1088/1755-1315/1044/1/011001

COMMITTEES

Conference Co-Chairs

Gordon Huang, University of Regina, Canada

Qudaih Yaser, Higher Colleage of Technology, UAE

Yongping Li, Beijing Normal University, China

International Program Committee

Chunjiang An, Concordia University, Canada

Sayed M. Bateni, University of Hawaii at Manoa, USA

David Campos-Gaona, University of Strathclyde, UK

Bing Chen, Memorial University, Canada

Huichao Chen, Southeast University, China

Yongsheng Chen, Chinese University of Hong Kong, China

Zhe Chen, Aalborg Universitet, Denmark

Hossam A. Gabbar, UOIT, Canada

Ahmad Z. A. Halim, Universiti Malaysia Pahang, Malaysia

Anthony Halog, University of Queensland, Australia

Svenja Hanson, University of Nottingham Ningbo, China

Guangwei Huang, Sophia University, Japan

Wendy Huang, University of Calgary, Canada

Aeslina A. Kadir, Universiti Tun Hussein Onn Malaysia, Malaysia

Jianbing Li, University of Northern British Columbia, Canada

Jianyong Li, Institute of Hydroecology, China

Zoe Li, McMaster University, Canada

Chunhua Liu, The City University of Hong Kong, Hong Kong, China

Hsiao-Kang Ma, National Taiwan University, Taiwan, ROC

Hassan Migdadi, American university of Madaba, Jordan

Siti N. B. Rahmat, Universiti Tun Hussein Onn Malaysia, Malaysia

Manousos Valyrakis, University of Glasgow, United Kingdom

Shuguang Wang, Shandong University, China

Xiaolin Wang, University of Tasmania, Australia

Zheng Xiao, Hunan University, China

Wenxian Yang, Newcastle University, United Kingdom

Hong Yue, University of Strathclyde, UK

Peer Review Statement

All papers published in this volume have been reviewed through processes administered by the Editors. Reviews were conducted by expert referees to the professional and scientific standards expected of a proceedings journal published by IOP Publishing.

- **Type of peer review:** Double Anonymous
- **Conference submission management system:** CMT System
- **Number of submissions received:** 32
- **Number of submissions sent for review:** 27
- **Number of submissions accepted:** 14
- **Acceptance Rate (Submissions Accepted / Submissions Received × 100):** 43.8
- **Average number of reviews per paper:** 2
- **Total number of reviewers involved:** 20
- **Contact person for queries:**
  Name: Con Cong
  Email: con_cong@apise.org
  Affiliation: Asia Pacific Institute of Science and Engineering

Research and Application of Bottom Water Control and Suppression Technology by CO_2 Huff-and-Puff

Tao Huang[1,a], Yanfeng Shao[2,b], Qingyu Liu[2,c], Li Li[1,d] and Xuewei Li [2,e,*]

[1] Oil production technology institute, Dagang Oilfield Company, China National Petroleum Corporation, Tianjin, China
[2] Oil and gas development department, Dagang Oilfield Company, China National Petroleum Corporation, Tianjin, China

[a]firevox@163.com; [b]shaoyanfeng2020@163.com; [c]liuqingyu2020@163.com; [d]dg_lili@petrochina.com.cn; [e,*]6c_lixuew@petrochina.com.cn

Abstract. In order to solve the problems about rapid increasement rate of water cut and large production decline in bottom water heavy oil reservoir due to active bottom water and high crude oil viscosity, the research on CO_2 huff-and-puff technology is carried out. The results showed that the mechanism of CO_2 huff-and-puff was mainly to reduce the viscosity of heavy crude oil, carbonate reacted with bottom water and formed interlayer to achieve viscosity reduction and precipitation. The comprehensive water cut was decreased by 35%, and the average oil production of a single well was increased by 3~4 times. Therefore, CO_2 huff-and-puff technology was one of technologies about bottom water control and oil production stabilization, and it is of great significance to enhance oil production in heavy oil bottom water reservoirs.

1. Introduction

Heavy oil is a kind of crude oil with high asphaltene and colloid content and high viscosity. In recent decades, the contribution of heavy oil to global oil production continues to grow and has become a research hotspot in recent years [1,2]. The traditional heavy oil exploitation method is mainly thermal recovery [3]. In 1981, Butler et al. first proposed steam-assisted gravity drainage technology (SAGD), which has developed into one of the most basic methods for heavy oil, especially extra-heavy and extra-heavy oil reservoirs, and achieved good field results [4]. However, conventional SAGD technology has some problems in actual production, such as large heat loss, large amount of water in the later stage, which leads to cost increase. In 1997, Butler formally proposed an improved SAGD technology - non-condensate gas plus (SAGP) [5], in which CO_2 is widely used.

To apply CO_2 to the recovery of heavy oil reservoirs, a lot of research has been carried out on CO_2 steam-assisted recovery of heavy oil reservoirs. In 2007, Sohrabi et al. applied numerical simulation methods to study the effects of CO_2 flooding, water flooding and CO_2 gas-water alternating flooding in heavy oil reservoirs and found that CO_2 gas-water alternating flooding increased recovery by about 30% compared with CO_2 flooding and water flooding alone [6]. In 2009, Tao et al. studied the mechanism of CO_2-assisted steam stimulation for ultra-heavy oil reservoir and found that the viscosity of ultra-heavy oil reservoir is very sensitive to temperature change, and the ultra-heavy oil reservoir expands after saturated with CO_2, which can increase the elastic energy of the reservoir [7]. In 2010, Yang et al. studied the physical properties of super heavy oil reservoirs and their interaction mechanism with CO_2 and found that the viscosity of crude oil in super heavy oil reservoirs decreased significantly after CO_2 was dissolved, and the greater the pressure, the greater the decrease [8]. In 2013, Li et al. studied the viscosity reduction mechanism and effect of CO_2 on heavy oil reservoirs and found that when CO_2 solubility reached 55 m^3/m^3, the viscosity reduction rate reached more than 95% [9].

The above studies on CO_2 flooding production of heavy oil reservoirs lacks quantitative evaluation of the solubility of deep-level CO_2 in ultra-heavy oil reservoirs and the influence of CO_2 on the density and viscosity of crude oil in ultra-heavy oil reservoirs. At the same time, due to the backward experimental equipment, limited by the sealing conditions, the heat resistance range of high temperature and high-pressure mixing barrel under high pressure is small, which needs to be studied.

2. Oilfield overview

K oilfield is a heavy oil bottom water reservoir, and the main development horizon is Guansan, and there are two wells in the K oilfield. The buried depth of the reservoir is 1200-1500m, the porosity and permeability are 35% and 1000-1500mD, respectively. It is a high porosity and high permeability reservoir with crude oil density of 0.975g/cm^3 and crude oil viscosity of 2380-3385 mPa·s. S oilfield is an ordinary heavy oil reservoir with pressure coefficient of 0.97 and oil reservoir temperature of 56 °C, which belongs to normally stable pressure system. Besides, Guantao formation belongs to fluvial facies, and subspecies are low curvature fluvial facies. The reservoir thickness is in a range between 10 and 20m, the thickness difference on the plane is large, thickness of bottom water is 35m, and the length of interface between crude oil and water is 1344m. The oil-bearing area is 1.1km^2 in K oilfield, geological reserves and a cumulative oil production was 1.2735 million tons and 216 thousand tons, respectively. Besides, the recovery degree of geological reserves is only 16.96%, but the water cut exceeds 90% after 24 months. At present, there are 15 production wells and 5 water injection wells in K oilfield, the daily oil production is only 23.39 tons, and comprehensive water cut is as high as 96.5%. Additionally, the bottom water conning problem was serious. Even if the injection formation layer was optimized and adjusted, the high water cut problem was not solved.

3. Study on Mechanism of CO_2 huff-and-puff

Aiming at the problem of high comprehensive water cut in heavy oil reservoir with active bottom water, in order to improve the development effect of K oilfield again, the pilot experiment of CO_2 huff-and-puff technology was carried out in April 2020. Until October 2021, the daily oil production of the oilfield has increased from 23.39 tons to 58 tons, and the comprehensive water cut has decreased from 96% to 89%.

3.1. Volume expansion of crude oil

The rapid breakthrough of water cut in bottom water reservoir is mainly due to the existence of dominant flow channel, resulting in bottom water coning and high water cut. By establishing the geological model of bottom water reservoir and carrying out numerical simulation, the results show that the volume of crude oil was expanded, and the formation water was discharged from the dominant channel of water flow after CO_2 was injected in the reservoir. It can be speculated that the oil-water barrier was established by carbonation of CO_2 and water. Besides, the oil phase separation flow and oil saturation in local well area is improved, which is drived by CO_2.

3.2. Extraction of heavy residual oil from crude oil

By analysing the physical properties of crude oil before and after CO_2 huff-and-puff in K oilfield, it is found that part of the heavy oil is diluted and dissolved after CO_2 injected into the reservoir, which improves the seepage capacity of heavy oil. Besides, the total salinity and the mass concentration of bicarbonate ion increase significantly after CO_2 huff-and-puff, indicating that CO_2 dissolves in water to form carbonic acid, which reduces the fluidity of water and improves the reservoir permeability, therefore production of single well is improved, as shown in Figure 1.

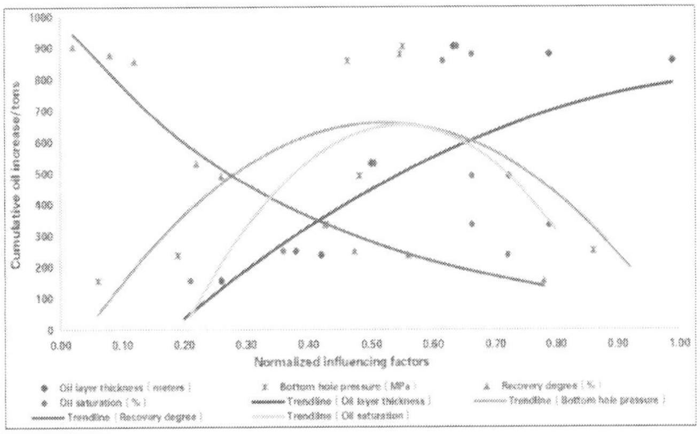

Figure 1. The chart of quantitive analyzation of influencing factors on CO_2 huff-and-puff

4. Establishment of well selection standard for CO_2 huff-and-puff

In order to promote the application of CO_2 huff-and-puff technology in heavy oil bottom water reservoir, the selection of suitable oil well is very important for improving the oil production. In view of many factors affecting CO_2 huff-and-puff degree in reservoir, through the investigation of implemented wells in Jidong Oilfield, combined with the practical experience of huff-and-puff in domestic heavy oil bottom water reservoir, the basic parameter database is established and calculated by different measurement methods. The criteria for well selection of heavy oil bottom water reservoir are established from 4 categories parameters, namely, reservoir geological characteristics, reservoir characteristics, fluid properties and development characteristics, which can provide reference for the implementation of huff-and-puff technology in heavy oil bottom water reservoir.

4.1. Formation pressure

Formation pressure is a necessary factor for CO_2 huff-and-puff. The dissolution effect of CO_2 in reservoir depends on the reservoir pressure, which improves the physical properties of crude oil [10-11]. Through the statistics of reservoir pressure and oil properties, it is found that there is an obvious relationship between oil production and formation pressure, so a total pressure range for CO_2 huff-and-puff needs to be determined.

4.2. Crude oil viscosity

The main mechanism of CO_2 huff-and-puff is dissolution, viscosity reduction, solubilization expansion and extraction. Part of the light oil has been recovered by CO_2 displacement, and most heavy oil remains underground. Through CO_2 huff-and-puff, the viscosity of the heavy oil is reduced, and the fluidity of the oil near the well is improved [12-13]. Finally, the heavy oil is displaced, and the water cut is greatly reduced.

4.3. Reservoir thickness

At present, the oil layer thickness of huff-and-puff wells in other countries is 16-67m, and the oil layer thickness of huff-and-puff wells in K oilfield is 5-15m. There is a positive correlation between the reservoir thickness and the oil production. The oil production by CO_2 huff-and-puff increases with the increase of reservoir thickness [14-15]. In addition, in order to reduce interlayer interference, the number of external reservoirs and well section span should be small.

4.4. Remaining oil saturation

The larger the remaining oil saturation can improve CO_2 huff-and-puff effect. From the perspective of economic benefit, the oil-rich reservoir in the initial stage is suitable for cold production with small investment [16-17]. CO_2 huff-and-puff production was started after high water cut in the later stage

[18-19]. The remaining oil saturation range was from 34 to 85% when CO_2 huff-and-puff was implemented abroad, and the original saturation of K oilfield was 64%. After a long period of development, the remaining oil saturation was predicted to be greater than 35% through numerical simulation, that is, the set critical saturation was greater than 35%.

4.5. Reservoir and well selection factors

All parameters including 11 wells and other fields are collected and normalized to form an effective chart in this work. Therefore, the criteria for reservoir and well selection can be determined, as shown in **Table 1**.

Table 1. Screening conditions of huff-and-puff wells.

Screening factors	Reservoir type	Crude oil viscosity	Permeability	Porosity	Reservoir thickness	Oil saturation
Range	Bottom water reservoir	2000-8000mPa·s	Above 0.9	27%	Above 4m	More than 35%

5. Application examples

Based on the selection standard of well and reservoir, CO_2 huff-and-puff technology of heavy oil bottom water reservoir was carried out. From October 2020 to September 2021, 28 wells were carried out, which increases by 85 tons in daily production and decreases by 20% in comprehensive water cut. Besides, increasement of cumulative oil is 12000 tons. In the implementation process of the project, the decline trend of CO_2 huff-and-puff wells is huge, which can be divided into four stages, namely, low water cut period, water cut rapid increase period, horizontal stability period and water cut slow increase period, as shown in Figure 2 and Figure 3.

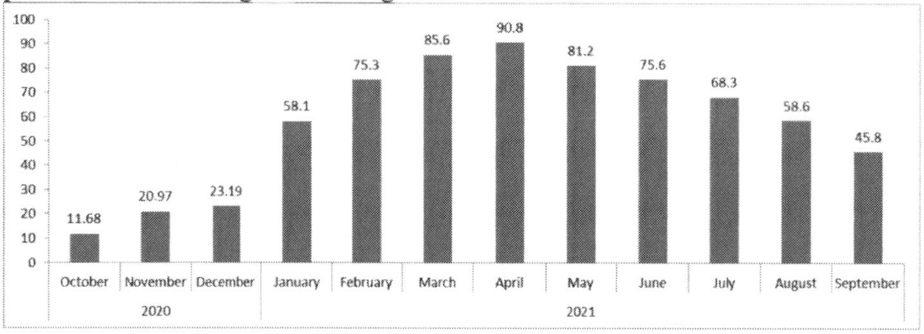

Figure 2. Daily oil production characteristics of implementation block

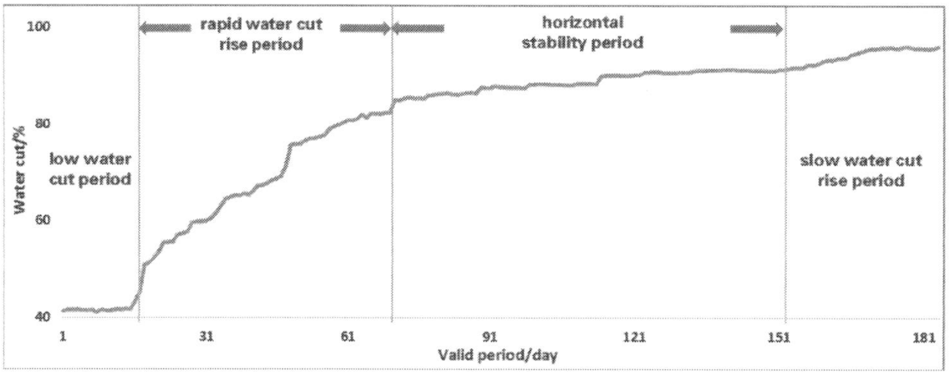

Figure 3. Water cut characteristics of implementation block

5.1. Low water cut period
The overall low water cut period of the K oilfield is about 17.8 days on average, and the low water cut period of conventional wells is 20 days. The existence of the low water cut period is due to the dissolution of CO_2 in the remaining oil in the formation, resulting in a viscosity reduction and expansion effect, which improves the oil phase separation mobility ratio. At the same time, the carbonation and Jia Min effect were generated by fusion of water and CO_2, which reduces the water phase separation. The seepage capacity of expanded crude oil was enhanced by mobility ratio, resulting in crude oil moving along the fluid domain channel instead of water.

5.2. Water cut rapid increase period
Because the production process destroys the original reservoir conditions including temperature and pressure, the CO_2 originally dissolved in the crude oil breaks through the large pores with advantageous migration capabilities, resulting in the rapid release of expansion energy, and the rapid rise of water content. The overall rapid water cut rising period of K oilfield is 53.2 days, in which the average water cut rising period of horizontal wells is about 60 days, and the average water cut rising period of conventional vertical wells is only 36.8 days. The differences between horizontal wells and conventional wells in this period is due to the larger contact area of CO_2 with fluids based on longer horizontal section of horizontal wells in the formation. Under the same conditions of CO_2 injection and simmering time, horizontal wells have higher CO_2 retention, better water control and oil stabilization effects, and a longer water cut rise period.

5.3. Horizontal stability period
As the degree of crude oil production, CO_2 huff-and-puff wells gradually enter the high water cut production period and begin to enter the late stage of CO_2 huff-and-puff. The heavy oil with far-reaching CO_2 is slowly carried out with the flow of bottom water, the overall production situation tends to be stable, and there is basically no intermittent occurrence. This period lasts for about 80 days on average. The production fluctuation of horizontal wells during this period is more stable than that of vertical wells, because the length of the horizontal section makes the liquid supply capacity stronger.

5.4. Water cut slow increase period
The water cut slow period of the high water cut stage lasts for about 30 days. During this stage, the water cut gradually increases until it reaches the level before the huff-and-puff. Most of the crude oil in the affected area is extracted, and the CO_2 huff-and-puff valid period ends here.

6. Conclusion
(1) The mechanism of CO_2 huff-and-puff in heavy oil bottom water reservoirs was mainly to reduce the viscosity of heavy crude oil, carbonate reacted with bottom water and formed interlayer to achieve viscosity reduction and precipitation.

(2) The comprehensive water cut was decreased by 35%, and the average oil production of a single well was increased by 3~4 times.

(3) The CO_2 huff-and-puff process in heavy oil bottom water reservoirs, which can be divided into four stages, namely, low water cut period, water cut rapid increase period, horizontal stability period and water cut slow increase period

References
[1] Lu C, Liu H, Zhao W, et al. Experimental investigation of in-situ emulsion formation to improve viscous-oil recovery in steam-injection process assisted by viscosity reducer [J]. SPE Journal, 2016, 22 (1): 130-136.
[2] Liu P, Zheng H, Wu G. Experimental study and application of steam flooding for horizontal well in ultraheavy oil reservoirs [J]. Journal of Energy Resources Technology, 2017, 139 (1): 1-9.

[3] Yu Liandong. Distribution of world heavy oil reserves and its re- covery technologies and future [J]. Special Oil & Gas Reservoirs, 2001, 8 (2): 98-103.
[4] Butler R M, Stephens D J. The gravity drainage of steam-heated heavy oil to parallel horizontal wells [J]. Journal of Canadian Petro- leum Technology, 1981, 20 (2): 36.
[5] Butler R. The steam and gas push (SAGP) [J]. Journal of Canadian Petroleum Technology, 1999, 38 (3): 54-61.
[6] Sohrabi M, Jamiolahmady M, Quraini A A. Heavy oil recovery by liquid CO_2/water injection [R]. SPE 107163, 2007: 1-8.
[7] Tao Lei, Li Zhaomin, Zhang Kai, et al. Study on the mechanism of CO_2-assisted steam puff and huff in ultra-heavy oil reservoirstaking west area of Zheng411, Wangzhuang Oilfield as an example [J]. Petroleum Geology and Recovery Efficiency, 2009, 16 (1): 51-54.
[8] Ouyang Chuanxiang, Du Xiaoxia.Physical property of superheavy oil and the mechanism analysis of the interaction with CO_2 [J]. Drilling & Production Technology, 2010, 33 (4): 90-93.
[9] Li Zhaomin, Sun Xiaona, Lu Teng, et al. Study on CO_2 mechanism in heavy oil thermal recovery for the Mao8 block [J]. Special Oil & Gas Reservoirs, 2013, 20 (5): 122-124.
[10] Hou Ganggang, Ma Xiaoli, Zhao Wenyue, Diwu Pengxiang, Liu Tongjing, Hou Jirui. Synergistic Modes and Enhanced Oil Recovery Mechanism of CO_2 Synergistic Huff-and-puff[J]. Energies, 2021, 14(12).
[11] Ding Mingchen, Wang Yefei, Liu Dexin, Wang Xia, Zhao Hailong, Chen Wuhua. Enhancing tight oil recovery using CO_2 huff-and-puff injection: An experimental study of the influencing factors[J]. Journal of Natural Gas Science and Engineering, 2021, 90.
[12] Guo Wenxuan, Zhao Renbao, Chen Changjian. Dissolution viscosity reduction characteristics and huff-and-puff effect of heavy oil in Chun 17 well area [J]. Journal of Xi'an University of Petroleum (NATURAL SCIENCE EDITION), 2021, 36(01):66-72.
[13] Tang Renxuan, Wu Gongyi, Chen Ju, Wang Jing, Wang Lianguo, Liu Juan. Analysis and understanding of influencing factors of CO_2 huff-and-puff effect in complex fault block reservoirs in Northern Jiangsu [J]. Energy at home and abroad, 2020, 25(12):32-38.
[14] Wu Xi, Zhang Zhuxin, Zhang Xiaoqing, Li Yunpeng, Chen Zixiang, Tang Yong. Heavy oil reservoir CO_2 in the middle and late stage of Dagang oilfield development optimization and practice of huff-and-puff parameters [J]. Oil and gas reservoir evaluation and development, 2020, 10 (03): 80-85.
[15] Sun Fengping, Liu Cunhui, Wang Zhihong, an Liyuan, Li leizhen. Application of huff-and-puff technology in Banqiao oilfield [J]. Chemical management, 2019(34):220-221+223.
[16] Che Zhengjia, Hu Zilong, Ma Li, Wang Shunli, Mei Rong, Cui Yinan. Influence of static parameters on CO_2 in Bannan fault block of Banqiao Oilfield study on the effect of huff-and-puff [J]. Offshore oil, 2019, 39(03):38-42.
[17] Jia Ying. Evaluation of CO2 huff-and-puff effect and analysis of influencing factors in qikeshu oilfield [J]. Progress of fine petrochemical industry, 2019, 20(04):15-18.
[18] Jin Chunyu, Xu Meixiang, Hu Chaoyang, Zhang Yonghui. Heavy Oil CO_2 in Nanpu Oilfield study on fuzzy comprehensive evaluation method of huff-and-puff well selection [J]. Practice and understanding of mathematics, 2019, 49(15):139-145.
[19] Zhang Tao, Li Dening, Cui Yinan, Liu Yonghe, Che Zhengjia, Wang Shunli. CO_2 in horizontal wells at ultra-high water cut stage in Banqiao oilfield optimization and implementation of huff-and-puff parameters [J]. Evaluation and development of oil and gas reservoirs, 2019, 9(03):51-56.

IOP Conf. Series: Earth and Environmental Science 1044 (2022) 012002 doi:10.1088/1755-1315/1044/1/012002

Biological Nitrogen Removal by A Membrane-Aerated Biofilm Reactor: An Attempt to Treat Urine Wastewater with High Sustainability

Chengbo Zhan[1,2,a], Liangchang Zhang[3,b], Weidang Ai[3,c] and Wenyi Dong[1*,d]

[1]Shenzhen Key Laboratory of Water Resource Utilization and Environmental Pollution Control, School of Civil and Environmental Engineering, Harbin Institute of Technology (Shenzhen), Shenzhen 518055, China
[2]Space Institute of Southern China, Shenzhen, 518117, China
[3]China Astronaut Research and Training Center, Beijing, 100094, China

[a]chengbo.zhan@gmail.com; [b]celsszhang@163.com; [c]aiwd@163.com;
[*d]dongwenyi@hit.deu.cn; dwy1967@qq.com

Abstract. Human urine contains the majority of nutrients in domestic wastewater, therefore treatment of source-separated urine by sustainable methods would become a low-cost and eco-friendly alternative to conventional urban wastewater treatment system. In this work, a waste stream composed of real urine and hygiene wastewater was prepared, which had characteristic of high nitrogen content and low available C/N ratio. Membrane-aerated biofilm reactor (MABR) is an emerging technology utilizing microporous or nonporous aeration membranes as the substrate for biofilm attachment, and this configuration could promote the growth of autotrophic nitrifying bacteria and the efficiency of nitrogen removal process. The waste stream was treated with a home-made MABR system to remove the organic carbon and nitrogen pollutants. Over a four-month operation period with the incoming waste stream concentration progressively elevated, the system achieved 90-95% removal of TOC, 50-73% oxidation of ammonium-nitrogen, and 24-63% removal of total nitrogen from the influent, without external carbon source input. The results indicated that MABR has the potential for urine wastewater treatment with low consumption in energy and materials.

1. Introduction

Stream segregation is a strategy to divide waste streams with different characterizations and then treat them separately, and by this approach, heavily polluted streams with low volume can be treated without contaminating high volume of lightly polluted streams or mixing contaminants from different streams, therefore the overall difficulties and cost of wastewater treatment would be reduced. For domestic wastewater, the waste streams can be categorized into grey water (wastewater from bathroom, laundry, and kitchen) and black water (wastewater from toilets), in which black water can be subcategorized into yellow water (urine and flushing water) and brown water (feces, toilet paper and flushing water)[1]. Grey water takes up a large portion of total volume for the domestic wastewater and is least polluted, yellow water is rich in nutrients (nitrogen, phosphorus, and potassium) and small molecule compounds (pharmaceutical products, hormones, and metabolites), while brown water has high concentration of organic matter and pathogenic microorganisms[2, 3]. Urine contributed about 80% of nitrogen and 50% of phosphorus loading of the total domestic waste stream with only 1% of the volume, so separated treatment of urine wastewater would have a significant impact on centralized wastewater treatment plants[4]. According to water access level provided by World Health Organization, for households water quantity of 100 L/person/day was optimal, and that of 20 L/person/day could only cover the basic need for food and hygiene[5]. Since the normal daily urine volume for an adult is 1-2 L, the ability to treat waste stream with the urine volume fraction of at least 10% would be meaningful for regions with only basic water adequacy and those with intermediate or optimal water adequacy.

The treatment of urine wastewater usually involves nutrient removal or nutrient recovery. Nutrient recovery orients production of fertilizers from urine wastewater, and emphasizes on the utilization of nutrients within wastewater[6], while nutrient removal cares about reducing the loading of the urine wastewater, focusing on improvement of water quality. From the energy perspective, nutrient removal by 90% of urine diversion could directly save 33% of aeration energy from wastewater treatment plant, according to a recent study.[7] On the other hand, nutrient removal could save the energy consumption by reducing the production of fertilizer, and the quantitative modeling would be more complicated, since more factors would be involved, including the nutrient recovery technology applied, the form of fertilizer recovered, and how this fertilizer is produced in industry. In practice, appropriate technologies should be employed to treat urine wastewater after consideration in economic cost and environmental impact. Physical technologies including thermal process and membrane technology are high in cost and energy consumption compared to biological treatment methods[8]. In average, membrane technology such as RO and NF has energy consumption of 0.79 - 0.92 kWh to treat per cubic meter of water; the energy consumption of thermal processes including multi-stage flash distillation and vapor compression distillation could be up to 7.9-68.69 kWh/m^3; the energy consumption for biological technology such as conventional activated sludge was 0.1 - 0.60 kWh/m^3[9]. Chemical technology such as struvite crystallization requires consistent reagent supply and the recovery may not be complete. As the formula of Struvite ($MgNH_4PO_4{\cdot}6H_2O$) indicates, the stoichiometry of Mg, N, and P are equivalent, whereas the molar ratios of Mg:N:P in typical urine are 1:50-100:10, the nitrogen content is highly excessive and magnesium is inadequate for this process[10]. After complete precipitation of phosphate, appropriate treatment for the majority of remaining N in the stream is required. Therefore, the biological technology has the advantages of high sustainability and low energy consumption.

Urine wastewater has characteristic of high nitrogen content and low C/N ratio, if this type of stream was treated by anaerobic-anoxic-oxic (AAO) process, the amount of aeration power and carbon source input would be tremendous to achieve complete nitrification and denitrification for each step. MABR is an emerging technology that utilizes the shell side of hydrophobic microporous or nonporous membranes as the substrate for biofilm attachment, whereas air enter the lumen side and diffuse across the membrane with the pressure gradient[11]. Therefore, the stratification of biofilm in MABR has a unique pattern that aerobic bacteria reside at bottom, which is reversed to the biofilms attached to a regular surface[12]. Since the carbon source has to pass the outer biofilm before delivered to this region, the growth of heterotrophic bacteria is inhibited and the efficiency of oxygen utilization for nitrification is optimized. The nitrite and nitrate products could be transported to outer layer biofilm, thus facilitating nitrogen removal processes such as denitrification, partial denitrification, or anaerobic ammonium oxidation (Annamox)[13]. The enhanced efficiency of utilization of oxygen and carbon source would reduce the energy and materials consumption, leading to more sustainable treatment of urine wastewater. In this work, a membrane-aerated biofilm reactor (MABR) was established to treat a synthetic stream constituted of real urine and sanitation wastewater, and the nutrient removal efficiency was evaluated. The reactor was fed with a stream with the urine volume fraction progressively elevated to 10%, and over a four-month period, it achieved 88-95% removal of TOC, 50-73% oxidation of ammonium-nitrogen, and 24-63% removal of total nitrogen from the influent. During the whole operation stage there was no external carbon source input. The results indicated MABR had high sustainability with low consumption in energy and materials. Future work will focus on the optimization of the MABR operation parameters and tests with waste streams with higher urine volume fraction or pure source-separated urine.

2. Experimental

2.1. Materials
The reagents used in the experiment (listed in Table 1) were purchased from Shanghai Macklin Biochemical Co., Ltd with AR grade or above. Hygiene products, including toothpaste (Lengsuanling, Dencare), shampoo (Rejoice, P&G), facial cleanser (Mentholatum), shower gel (Safeguard, P&G), hand soap and laundry detergent (Blue Moon Industrial Co. Ltd), were purchased from JD.com. Urine was collected from one male bathroom located at the institute building all weekdays during the project, and the collected urine was stored at 4 °C before use. The nutrient media used for sludge activation and the influent stream used for acclimation and working stage was prepared according to the composition listed in Table 1. For nutrient media, a solution of concentrated micronutrients was prepared and then added to nutrient media. For influent stream, each batch was about 24 L in volume, and during preparation the hygiene products were weighed and dissolved first, followed by the addition of urine and water, and the stream was stored at 4 °C and equalized for at least 12 h before added to the influent tank. The solvent for all steps involved in nutrient media and influent stream preparation was tap water.

Table 1. Composition of nutrient media and influent waste stream. The concentration unit for each component listed is g/L for solids and mL/L for liquids

Nutrient media				Waste stream	
Nutrient media		Concentrated micronutrients			
$(NH_4)_2SO_4$	0.5	$CaCl_2 \cdot 2H_2O$	0.50	Toothpaste	1.0
$NaHCO_3$	1.0	$CoCl_2 \cdot 6H_2O$	0.35	Shampoo	1.0
KH_2PO_4	1.0	$CuSO_4 \cdot 5H_2O$	0.15	Facial cleanser	1.0
glucose	0.2	$FeCl_3$	0.80	Shower gel	1.0
Concentrated micronutrients	3	$MgSO_4 \cdot 7H_2O$	0.34	Laundry detergent	1.0
		$MnCl_2 \cdot 4H_2O$	0.50	Hand wash	1.5
		$Na_2MoO_4 \cdot 2H_2O$	0.20	Urine	100
		$ZnSO_4 \cdot 5H_2O$	0.55		
		Yeast powder	10		

2.2. MABR setup
An MABR module was fabricated for the purpose of this experiment. The module was consist of 1440 PVDF hollow fibers with diameter of 1.9 mm, acrylic shell, and epoxy resin sealing. Both distal ends of the fibers were left open and the gas flow had flow-through configuration. A pressure gauge was set before the inlet to monitor the aeration pressure and the gas flow rate of inlet air and exhaust air was

measured by two glass rotameters. The length contacting the bulk liquid for the fibers was approximately 16 cm, and the overall surface area of the hollow fibers was about 1.37 m^2.

After the module was fabricated, the MABR was setup as Figure 1. Two submerged pumps were put at the bottom of the reactor tank, subsequently the MABR module was immersed in the reactor tank. The influent stream was transferred from the influent tank continuously by a peristaltic pump, and the excessive liquid would overflow to the effulent tank when the liquid level reached an opening on the wall of the reactor tank. During operation, the submerged pumps were kept working to ensure the reaction liquor was thoroughly mixed.

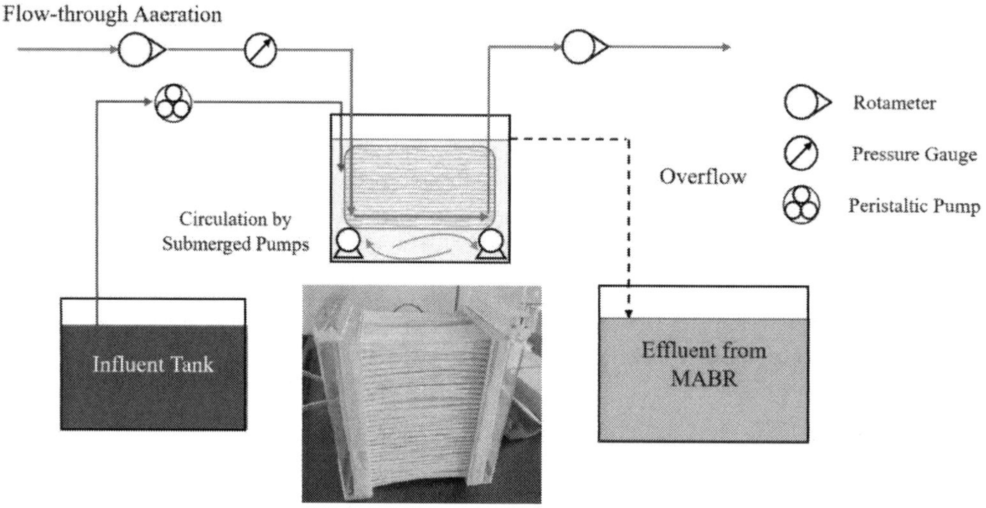

Figure 1. The configuration of the MABR. The inset at the bottom was a picture of the MABR module before the operation of the system.

2.3. Methods for water quality assessment

During the experiment period, aliquots were taken from influent tank and effluent tank about 1 time every two days, and the TOC was measured by a Shimazu-TOC-LCPH analyzer (Shimazu) with automatic sampler. Total nitrogen (TN) for influent aliquots was determined by alkaline persulfate digestion method. The ammoniacal nitrogen (NH_3-N), nitrite-nitrogen (NO_2-N) and nitrate-nitrogen (NO_3-N) of effluent aliquots were measured to analyze the nitrogen conversion. NH_3-N was measured by a Flowsys III continuous flow analyzer (Systea) with salicylate method. NO_2-N and NO_3-N were measured by a Dionex-ICS 5000+ ion chromatography system (Thermo Fisher Scientific) equipped with an AS11-HC column for anion exchange.

3. Results and Discussion

3.1 Initiation and acclimation of the MABR system

After the MABR system was setup, activated sludge (acquired from return sludge tank of secondary sediment tank at Hengling Wasterwater Treatment Plant, Shenzhen, China) was inoculated to the system. The system was fed with nutrient media for one week before the influent was switched to waste stream. Meanwhile, the aeration was initiated and throughout the experiment period, the aeration pressure was kept in the range of 4.5-6 kPa.

After inoculation of activated sludge, the MABR system was fed with synthetic nutrient media for one week. Then, to acclimate the sludge, the volume fraction of the waste stream was progressively elevated from 0% to 100% through 6 stages: stage 0 (Day 1-7): nutrient media; stage 1 (Day 8-20): 10% waste stream + 90% tap water; stage 2 (Day 21-37): 25% waste stream + 75% tap water; stage 3 (Day 38-61): 40% waste stream + 60% tap water; stage 4 (Day 62-88): 70% waste stream + 30% tap water; stage 5 (Day 89-117): 100% waste stream. The hydraulic retention time (HRT) was 2.25 days for stage 0-3 , and 2.99 days for stage 4-5.

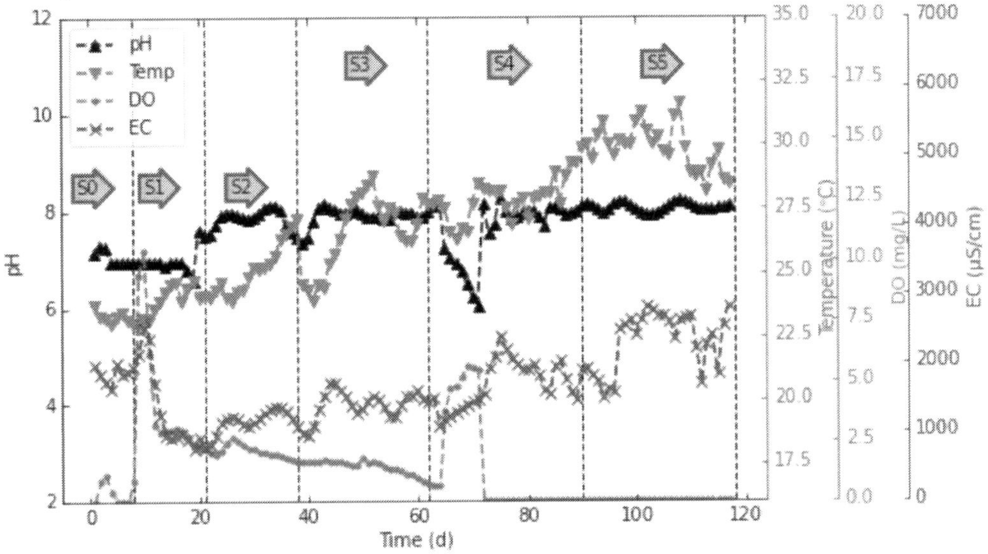

Figure 2. The pH, Temperature, DO, and EC profiles of the MABR system during stage 0-5 (S0-S5). The time frames were indicated by the vertical dot lines.

The profiles for pH, temperature, dissolved oxygen (DO), electrical conductivity (EC) of the MABR system through all stages were shown in Figure 2. In general, the pH stayed within the range of 7-8, and the temperature was within the range of 23-30 °C, which was appropriate for a good microbial activity. The only exception was for day 65-72 when the operation was suspended due to a travel plan, and the influent was stopped, leaving the aeration to maintain the microbial activity. Therefore, there was a boost of DO and a drop of pH, which was turned back after the operation was resumed. The DO exhibited a declining trend as the volume fraction of the waste stream elevated, since the consumption of oxygen increased as with the substrate concentration for nitrification and organic pollutent degradation. By the end of S4, the DO was extremely low and the bulk liquid phase of the MABR was at anoxic state. The EC was gradually elevated as volume fraction of urine elevated, which was reasonable because most of the ions was not degradable or assimilable by the sludge, such as Na^+ and Cl^-. Throughout the operational stages, the physicochemical properties were maintained at an appropriate level for the microbial community.

3.2 TOC removal efficiency

The TOC of influent and effluent was measured every 2-3 days since S1 to determine the TOC removal efficiency of the MABR system. The results were shown in Figure 3, including the mean value and standard deviation of influent TOC and effluent TOC for each stage. The removal efficiency was calculated using the difference between influent and effluent TOC divided by the influent TOC. Whereas most of TOC was originated from the urine and there was fluctuation of the TOC concentration (the deviation was big compared to the average value), the average influent TOC concentration was progressively rising as with the volume fraction of waste stream. Meanwhile, for all stages, the average effluent TOC concentration was significantly lower than corresponding influent TOC, indicating high TOC removal efficiency.

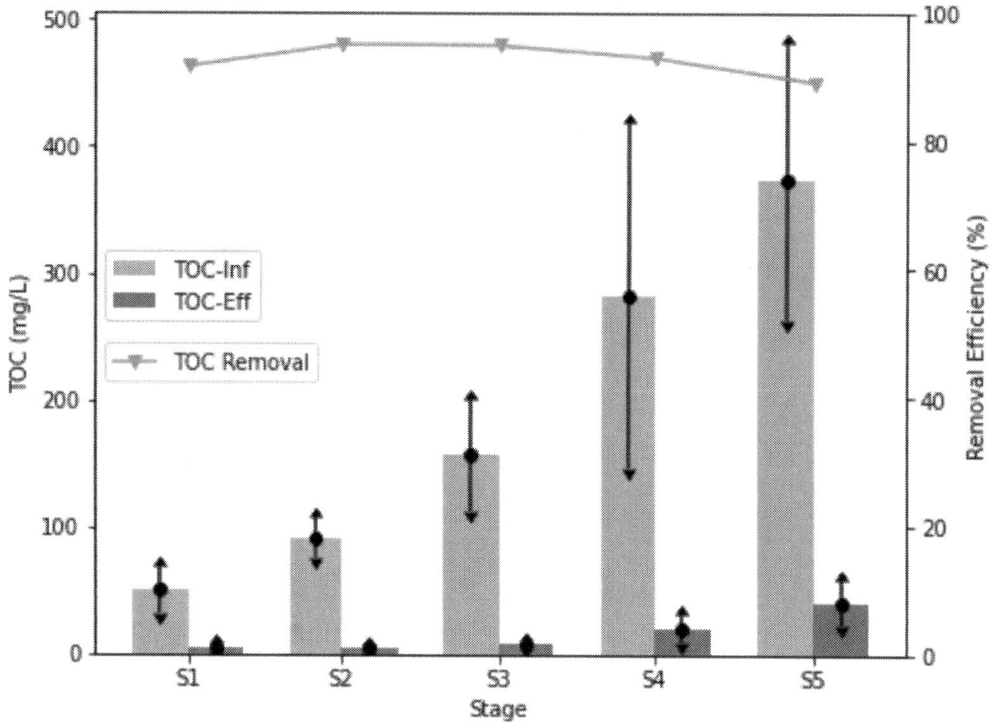

Figure 3. The TOC profiles and TOC removal efficiency during stage 1-5 (S1-S5). The bars indicated the mean values for influent and effluent TOC concentrations, and the black error bars were the deviations corresponding to the mean values.

It was shown in Figure 3 that the removal efficiency of TOC was high for all stages. As the volume fraction of waste stream increased from 10% to 100%, the average influent TOC concentration was 50.3, 91.8, 157.1, 282.9 and 373.2 mg/L for S1-S5, respectively. The average effluent TOC concentration for these stages was 4.27, 4.70, 8.22, 20.3, and 40.8 mg/L. Therefore, the average removal efficiency of influent TOC was around 90-95% (shown in Table 2) for each stage. Since the TOC loading was climbing up, there was a lowering of TOC removal efficiency and an escalation in average effluent TOC concentration. However, the MABR could remove at least about 90% of TOC, indicating the acclimation was successful.

During S5, an aliquot was taken from influent and effluent to test the concentration of linear alkylbenzene sulfonate (LAS). This test was meant for evaluation on the removal of anionic surfactant in the waste stream, which was brought by the hygiene products. The result was that the effluent LAS concentration was < 0.05 mg/L, whereas the influent LAS concentration was 12.8 mg/L. The removal of LAS was > 99%, meaning that the surfactant components could be effectively removed by the MABR after the acclimation.

3.3 Nitrification and nitrogen removal performance

The profiles for TN in influent, TN in effluent, and nitrogenous components (NH_3-N, and NO_x-N) in effluent were demonstrated in Figure 4. The nitrification efficiency was calculated from the influent TN and effluent NH_3-N, and the denitrification efficiency was calculated from the influent TN and effluent TN. As with the case of TOC, there was fluctuation of the influent TN concentration, since most of the nitrogenous pollutant was contributed by urine. Also, the average influent TN concentration was progressively rising as with the volume fraction of waste stream. For all stages, the average effluent TN concentration was lower than corresponding influent TN, but the removal efficiency was not as high as that for TOC.

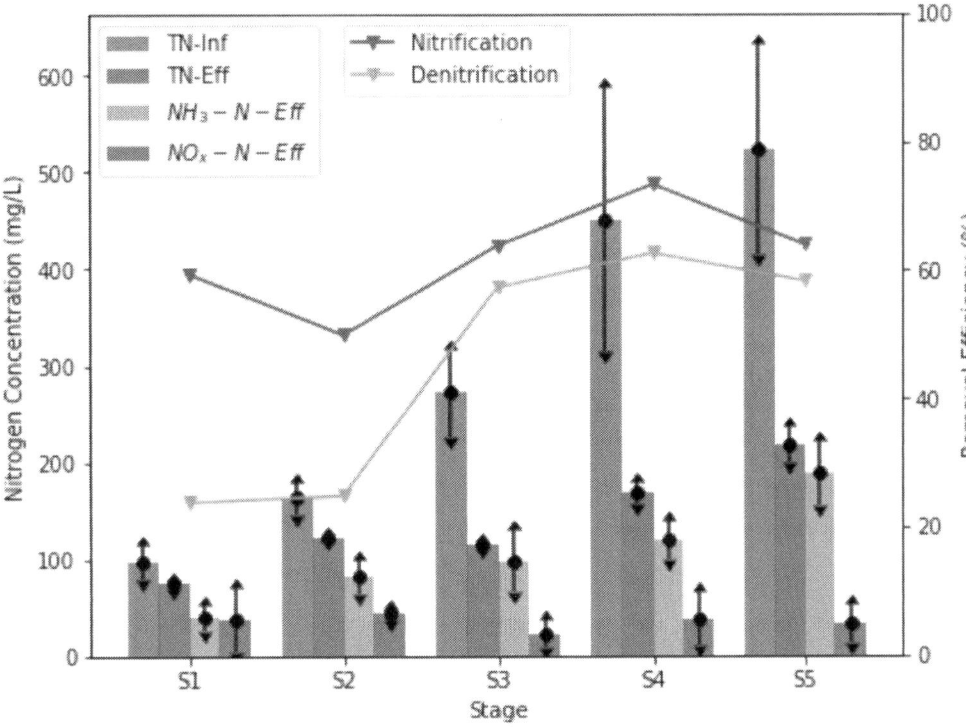

Figure 4. The profiles for nitrogenous pollutant (including TN, NH_3-N, and NO_x-N) and efficiency for nitrification and denitrification during stage 1-5 (S1-S5). The bars indicated the mean values for influent and effluent nitrogenous pollutant concentrations, and the black error bars were the deviations corresponding to the mean values.

As the volume fraction of waste stream increased from 10% to 100%, the average influent TN concentration was 97.6, 163.0, 271.6, 450.3 and 522.8 mg/L for S1-S5, respectively. The average effluent TN concentration for these stages was 74.3, 122.4, 115.7, 168.0, and 217.8 mg/L. As shown in Table 2, the denitrification efficiency was about 20% at S1 and S2, then rose to around 60% since S3. In the mean time, the nitrification efficiency stayed around 60-70%. The removal efficiency for nitrogenous pollutant was lower than that for TOC was probably that the degradation of organic compounds was mostly achieved by heterotrophs, but nitrification was mostly achieved by autotrophs. In general, the growth rate of heterotrophs were much higher, therefore, the removal efficiency for TOC reached a high level in a short time. However, the growth rate of ammonia oxidizing bacteria was limited by its autotrophic nature, therefore the nitrification efficiency was moderate. As for denitrification process, since most denitrifying bacteria was heterotrophic, their growth rate should be high. However, the substrates involved in this process (NO_x^-) were products of nitrification process, so the denitrification efficiency could not exceed nitrification in theory. On the other hand, denitrification was promoted in anoxic envirionment where NO_x^- were favorable electron acceptors; for S1 and S2, the DO in the system was high until S3 the environment became anoxic. Therefore, the denitrification efficiency was low at first, then exhibited a trend to increase. The results for nitrogen species in effluent suggested that the majority of effluent nitrogen was NH_3-N, and only small amount of NO_x^- existed. This fact also supported that denitrification was limited by nitrification, as NO_x^- were consumed while NH_3-N was excessive.

Table 2. Summary of removal efficiency for TOC and nitrogenous compounds in each stage

Stage	TOC removal (%)	Nitrification (%)	Denitrification (%)

S1	91.5	59.6	23.9
S2	94.9	49.9	24.9
S3	94.8	63.9	57.4
S4	92.8	73.3	62.6
S5	89.1	64.0	58.3

3.4 Microbial community composition

After acclimation, a sludge sample was acquired from the reactor for DNA sequencing. The microbial community composition was exhibited in Figure 5 at phylum level and genus level. The composition at phylum level exhibited most microorganisms of the sludge were of the phylum proteobacteria. The composition at genus level exhibited the 20 genera with the most abundance, in which *Nitrosomonas* was assigned as ammonia oxidizing bacteria, while most listed genera could be assigned as denitrification bacteria, such as *Acidovorax*, *Alicycliphilus*, *Pseudomonas*, *Thauera*, *Comamonas*, *Hydrogenophaga*, *Hyphomicrobium*, and *Methyloversatilis*. *Sphingopyxis* and *Novosphingobium* had the potential to degrade aromatic compounds, which could help the complete degradation of the pollutants in the waste stream. The high abundance of ammonia oxidizing bacteria and denitrification bacteria in microbial community composition accounted for the high removal efficiency of TOC and TN.

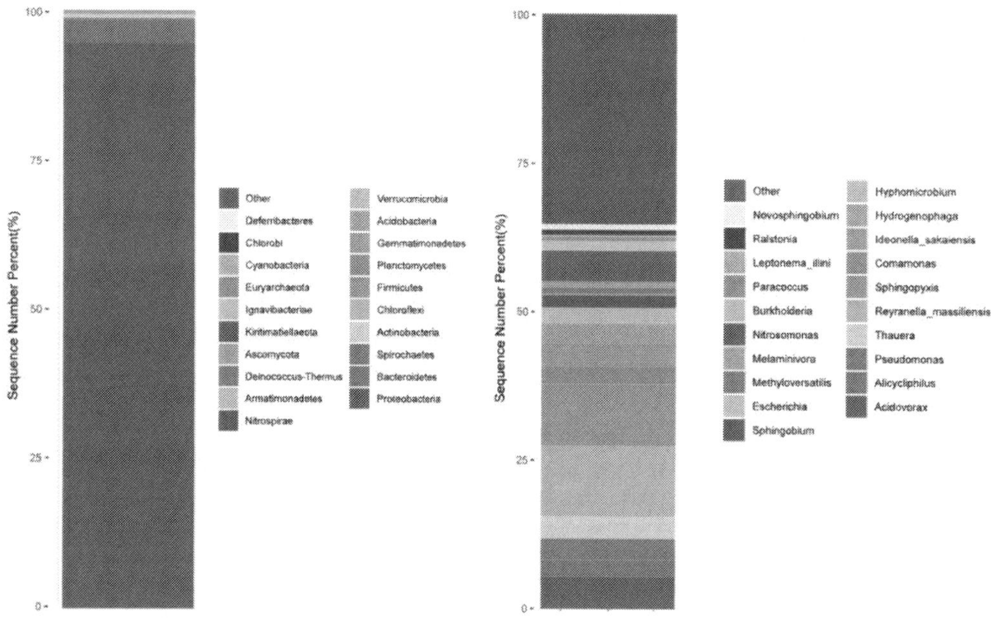

Figure 5. The microbial community composition at phylum level (left) and genus level (right).

4. Conclusions

In this work, a denitrification system was established based upon a home-made MABR module. After acclimation, waste stream with 10% volume fraction of urine could be directly fed into the reactor, and high removal of TOC and LAS was achieved. Despite nitrification and subsequent denitrification was a slower process than TOC degradation, the denitrification rate gradually went up, and when full strength waste stream was fed, about 60% of TN was removed. Also, the microbial community of the sludge demonstrated a variety of denitrifying bacteria. The facts confirmed the configuration of the MABR module, together with the operation parameters of the system, promoted the simultaneous nitrification-denitrification process, and the TN removal efficiency was probably limited by the nitrification rate and the limited amount of available carbon sources. This work provides insights of utilizing MABR technology to treat urine wastewater with low operation cost to reduce organic and nitrogen pollution.

Acknowledgements
This research was financially supported by National Key Laboratory of Human Factors Engineering, (Grant NO. 6142222200710 and SYFD061908). We sincerely thank Mr. Xiangfeng Cao, Mr. Yantai Chen, Ms. Mengting Hu, Mr. Weipeng Yang, and Mr. Shanjun Yin for their assistance in water quality tests. We also thank Wekemo Tech Group Co. Ltd (Shenzhen) for their genome sequencing and analysis service.

Reference

[1] Otterpohl R, Braun U and Oldenburg M 2004 Innovative technologies for decentralised water-, wastewater and biowaste management in urban and peri-urban areas Water Sci. Technol. 48 23-32
[2] Udert K, Larsen T A and Gujer W 2006 Fate of major compounds in source-separated urine Water Sci. Technol. 54 413-420
[3] Friedler E, Brown D M and Butler D 1996 A study of WC derived sewer solids Water Sci. Technol. 33 17-24
[4] Meinzinger F and Oldenburg M 2009 Characteristics of source-separated household wastewater flows: a statistical assessment Water Sci. Technol. 59 1785-1791
[5] Howard G, Bartram J, Williams A, Overbo A and Geere J-A Domestic water quantity, service level and health. 2nd ed. Geneva (Switzerland): World Health Organization 2020
[6] Verstraete W, Van de Caveye P and Diamantis V 2009 Maximum use of resources present in domestic "used water" Bioresour. Technol. 100 5537-5545
[7] Badeti U, Pathak N K, Volpin F, Dorji U, Freguia S, Shon H K and Phuntsho S 2021 Impact of source-separation of urine on effluent quality, energy consumption and greenhouse gas emissions of a decentralized wastewater treatment plant Process Saf. Environ. Prot. 150 298-304
[8] Ledezma P, Kuntke P, Buisman C J, Keller J and Freguia S 2015 Source-separated urine opens golden opportunities for microbial electrochemical technologies Trends Biotechnol. 33 214-220
[9] Wakeel M, Chen B, Hayat T, Alsaedi A and Ahmad B 2016 Energy consumption for water use cycles in different countries: A review Applied Energy 178 868-885
[10] Jönsson H and Vinnerås B Closing the loop: Recycling nutrients to agriculture. Source separation and decentralization for wastewater management. London (UK): IWA Publishing 2013: 163-178
[11] Terada A, Yamamoto T, Igarashi R, Tsuneda S and Hirata A 2006 Feasibility of a membrane-aerated biofilm reactor to achieve controllable nitrification Biochem. Eng. J. 28 123-130
[12] Martin K J and Nerenberg R 2012 The membrane biofilm reactor (MBfR) for water and wastewater treatment: principles, applications, and recent developments Bioresour. Technol. 122 83-94
[13] Gong Z, Liu S, Yang F, Bao H and Furukawa K 2008 Characterization of functional microbial community in a membrane-aerated biofilm reactor operated for completely autotrophic nitrogen removal Bioresour. Technol. 99 2749-2756

IOP Conf. Series: Earth and Environmental Science 1044 (2022) 012003 doi:10.1088/1755-1315/1044/1/012003

Exploration and Comparison of the Effects of Vibration and Backwashing Methods on the Process of the Sulfur-Based Autotrophic Denitrification

Baorui Liang[1,*,a], Tong Zhu[1,b], Fei Kang[1,c], Yao Wang[1,d] and Sai Yao[1,e]

[1]Institute of Process Equipment and Environmental Engineering, School of Mechanical Engineering and Automation, Northeastern University, Shenyang 110004, PR China

E-mail :[*,a] 1810136@stu.neu.edu.cn; [b] tongzhu@mail.neu.edu.cn; [c] 1747496195@qq.com; [d] 842116410@qq.com; [e] 2496840891@qq.com

Abstract. Both the vibration and backwashing methods are efficient pathway to improve the denitrification ability in reactors. However, the comparison of the effects of vibration and backwashing on the sulfur-based autotrophic denitrification (SAD) reactor is still lacking. In this study, two SAD reactors were compared under various conditions, and the results showed that the use of the oyster shell powder could lower the sulfate productivity with better denitrification ability, and the denitrifiers could be more enriched due to the use of the oyster shell powder. In addition, vibration method performs better than backwashing in improving SAD denitrification ability, reasonable vibration intensity can not only help to reduce energy consumption, but also improve the ability of denitrification of the SAD reactors to a greater extent. Overall, the results of the experiments in this study are helpful for improving the ability of SAD reactors at low temperature and could provide the possibility for wider application of SAD technology.

1. Introduction

Nitrate is a global pollutant and frequently found in shallow groundwater, in saliva, it can cause nitrosamine if the conversion of nitrate to nitrite, thus threatening the human health [1]; therefore, removing nitrate in nitrate-contaminated water is highly necessitated.

Biological processes are widely used for nitrate treatment in water bodies because they are more economical and efficient than other methods [2]. For the biological method, the heterotrophic denitrification (HD) process has been widely used due to its high efficiency. However, when the carbon/nitrogen (C/N) ratio is insufficient to support the denitrification process, the right dosage of water-soluble organic carbon sources is required, while it is difficult to control because of the complexity of the influent. From this point of view, the packed bed reactor was proposed because it is a cost-efficient method and not hard to operate. For the packed bed reactor, the sulfur-based autotrophic denitrification (SAD) is a preferred pathway and has been successfully studied in previous research [3]. The SAD process is an effective alternative to the HD process because it also has a strong denitrification ability. Nonetheless, the low temperature (≤15 °C) will affect the denitrification ability of the SAD process, which limits its application [4]. To alleviate the adverse effects of low temperature on the SAD process, scholars have done a lot of work, and the vibration and backwashing methods have been successfully demonstrated to enhance the denitrification ability of SAD at low temperatures [5,6], while few researches have focused on the vibration and backwashing methods, such as the influence of vibration intensity under the intermittent mode and backwashing intensity on the SAD process, as well as the comparison of the improvement of denitrification ability between them, has not been studied.

In this study, two kinds of the SAD filters were studied in two pilot-scale packed bed reactors. The overall aims of the study were as follows: 1) to compare the overall performance of the two kinds of SAD filters; 2) to investigate the effect of vibration intensity of intermittent vibration mode on the SAD process; 3) to study the effect of backwashing intensity on the SAD process; 4) to compare the variations in microbial communities of the SAD reactors. The result in this paper may provide help for the improvement of denitrification performance of SAD process at low temperature.

2. Materials and methods

2.1 The SAD filters and reactors

Two kinds of the SAD filters were prepared. The SAD filters were in the oval shape with the average particle size range was 6-8 cm^3. For the filters F_1 and F_2, under the weight ratio of 2:1, the elemental sulfur powders were mixed with oyster shell powder or calcium carbonate powder, and the mixed powders were respectively heated under 160 °C to produce the molten material, then the molten materials were respectively cooled in a water tank to make the filters F_1 and F_2. The diameter of the calcium carbonate powder, sulfur powder and oyster shell powder was 1200, 800, and 600 mesh, respectively.

Two identical up-flow reactors (R_1 and R_2) were prepared for the study. The reactor was column shaped, with an effective volume of 60 L (height: 80 cm, diameter: 31 cm). The reactors R_1 and R_2 were respectively filled the filters F_1 and F_2, and the weight of each kind of filter is 60 kg.

2.2 The configuration of the vibration and backwashing devices

A round plate (diameter: 26 cm) and 16 stainless steel rods were used to form the vibration device, eight motors (Leiming, China, 16 V DC motor, Maximum-4000 rpm/min) were equipped on top of the round plate and the motor speed was adjusted by a controller to adjust the intensity of vibration. A net which made of plastic was applied to wrap the rods to form a container.

For the backwashing device, the air pumps and the submersible pumps were equipped on the bottom of the reactors, and the backwashing intensity was adjusted by the pumps.

2.3 Experimental procedure

The start-up strategy and the synthetic wastewater for the reactors R_1 and R_2 were the same as the R_4-R_5 reactors in our previous study [6]. The detailed operating conditions of the R_1 and R_2 reactors are listed in Table 1. During periods 1 to 4, in order to compare the denitrification performance of the F_1 and F_2 filters, experiments were carried out under different conditions. During period 5, the temperature inside the reactors was maintained at 10 °C, to investigate the effect of intermittent vibration intensity and backwashing intensity on the SAD denitrification performance, the comparative experiments were carried out in 13 steps (Table 2). The calculation method of the hydraulic retention time (HRT) was the same as the previous study [2], and the flow rate of effluent is measured regularly to ensure the accuracy of HRT. During periods 1 to 4, the concentration of influent dissolved oxygen (DO) was less than 0.2 mg L^{-1}, and the DO was not controlled during period 5.

Table 1. Influent parameters of R_1 and R_2 reactors in this study.

Periods	1	2	3	4	5

Days	0-5	6-20	21-40	41-60	61-190
HRT (h)	3	2	1	0.5	3
NO_3^--N (mg L^{-1})			30		60
Temperature (°C)			29		10
DO of the influent (mg L^{-1})			0.2		2.1-4.0

During period 5, the filters in the reactors were taken out and respectively filled into the containers of the two same vibration devices, after the vibration device was respectively placed into the reactors, then the rest filter was filled back into the reactor. The vibration operates at three speeds, the motor rotational speed (rpm) of 2000, 3000 and 4000 corresponds to vibration intensity level 1, level 2 and level 3 respectively, under three vibration intensity levels, the vibration duration time is five seconds per minute to conduct intermittent vibration experiments. For the backwashing experiments, the operation was performed every five days, the duration time of the air flushing and water flushing was 12 min, and the detailed backwashing intensity is shown in Table 2.

Table 2. Operation conditions of R_1 and R_2 reactors in period 5.

Steps	1	2	3	4	5	6	7	8	9	10	11	12	13
Days	61-70	71-80	81-90	91-100	101-110	111-120	121-130	131-140	141-150	151-160	161-170	171-180	181-190
Mode (V:vibration; B:backwashing)	V	B	V	B	V	B	V	B	V	B	V	B	/
Vibration intensity	level 1	/	level 2	/	level 3	/	level 1	/	level 2	/	level 3	/	/
Backwashing intensity (A:air flushing (L·(s·m^2)$^{-1}$); W: water flushing (L·(s·m^2)$^{-1}$))	/	A:10; W:5	/	A:6; W:3	/	A:3; W:2	/	A:10; W:5	/	A:6; W:3	/	A:3; W:2	/
HRT (h)							3						
NO_3^--N (mg L^{-1})							60						

2.4. Analytical methods and DNA extraction and Illumina MiSeq sequencing

The influent and effluent samples of the reactors were daily collected and analyzed for ammonium, nitrite, nitrate, sulfate, alkalinity and mixed liquor suspended solids (MLSS). The methods of measurement are based on the Liang et al. [5], the DO and pH were detected according to Liang et al. [6].

On day 190, one bio-sample was obtained from the R_1 and R_2 reactors, respectively, and the DNA extraction and Illumina MiSeq sequencing methods were the same as Liang et al. [6].

3. Results and Discussion

3.1 Performance of R_1 and R_2 reactors

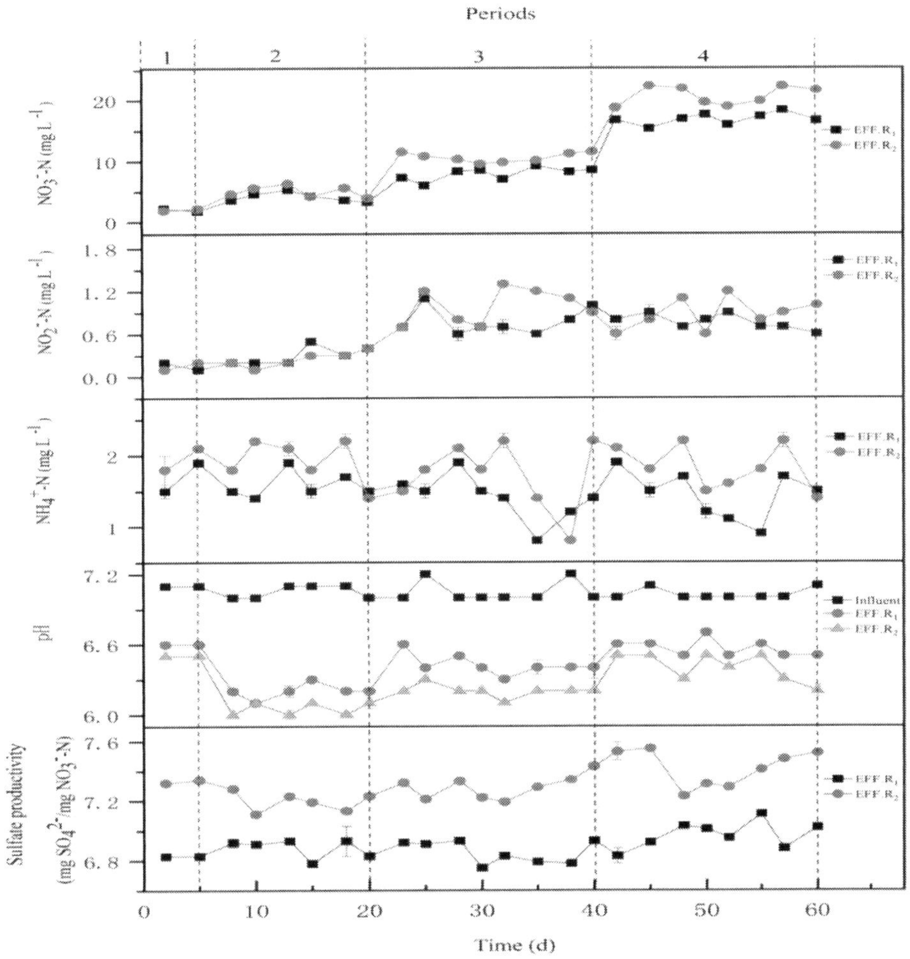

Figure 1. The overall performance of the R_1 and R_2 reactors from periods 1 to 4.

During periods 1 to 4, the performance of the R_1 and R_2 reactors is summarized in figure 1. From the periods 1 to 2, the nitrate removal efficiency (NRE) of the reactor R_1 was always higher than 80%, and R_1 reactor achieved better denitrification performance than R_2. When the HRT was $\leq$ 1 h in periods 3 to 4, the NRE in the reactors has dropped significantly, especially in period 4, the NRE of the R_1 and R_2 reactors was 43.9% and 29.6%, respectively. The reason for this result could be due to that the flow rate in the reactor is too fast, and the solubility of filters limits the NRE at fast flow rates. Compared with the reactor R_2, the reactor R_1 obtained lower sulfate productivity with higher pH rate in the effluent (figure 1).

In reactor R_1, the better denitrification performance and the lower sulfate productivity may be attributed to the use of oyster shell powder in filter F_1, because oyster shell powder contains a certain amount of organic carbon, which contributes to the HD process [7], thereby reducing the sulfate productivity, while the structure of oyster shell powder may be more conducive to the attachment of microorganisms [8], so it helps to enhance the denitrification capacity of the reactor. Overall, the reactor R_1 performed better than the reactor R_2, and the use of oyster shell powder in the SAD filter may be more conducive to the denitrification process.

3.2 Effects of vibration and backwashing methods on the SAD process

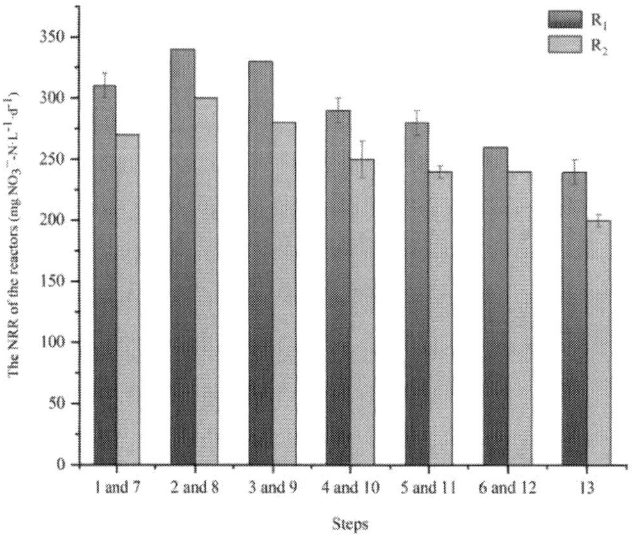

Figure 2. The NRR of the reactors in period 5.

During period 5, from the steps 1 to 12, for the vibration or backwashing method, the nitrate removal rate (NRR) at different steps but with the same intensity was averaged, and the NRR of the R_1 and R_2 reactors is depicted in figure 2. During the blank comparison condition (step 13), the average NRR of the R_1 and R_2 reactors was 241 ± 8.9 and 201 ± 4.1 mg NO_3^--N·L^{-1}·d^{-1}, respectively. Improvements in NRR were achieved at three different vibration intensities, specially, the NRR increased more obviously when the vibration intensity is level 2. Although the average NRR in steps 3 and 9 was higher than in steps 1 and 7, the concentrations of the MLSS in the R_1 and R_2 reactors effluent at level 2 were both higher than 10 mg L^{-1}. This indicates that high vibration intensity is of limited help to denitrification in SAD process, and may cause high concentration of suspended solids in effluent. In general, the vibration method helps to improve the denitrification performance of SAD process at low temperature, and the intermittent vibration method is feasible, while the specific vibration intensity should be controlled within a suitable range.

At three different intensities, the backwashing method also improved the denitrification performance in reactors. Excessive sludge in the reactor may affect the performance of denitrification [9], and the improvement of denitrification performance in reactors by backwashing may be due to the removal of aged sludge. And it is worth noting that, when the intensity of water flushing and air flushing was 5 and 10 L·(s·m^2)$^{-1}$, respectively, there will be SAD filters washed away, which may be detrimental to the denitrification process. However, in steps 6 and 12, lower NRR rate was obtained, which may be due to insufficient intensity of backwashing. Compared with the backwashing, the vibration method is more beneficial to the denitrification performance of SAD process (figure 2), this may be related to that vibration contributes to the solubility of SAD filter [5], thus improving the denitrification ability.

3.3 Evolution of microbial community

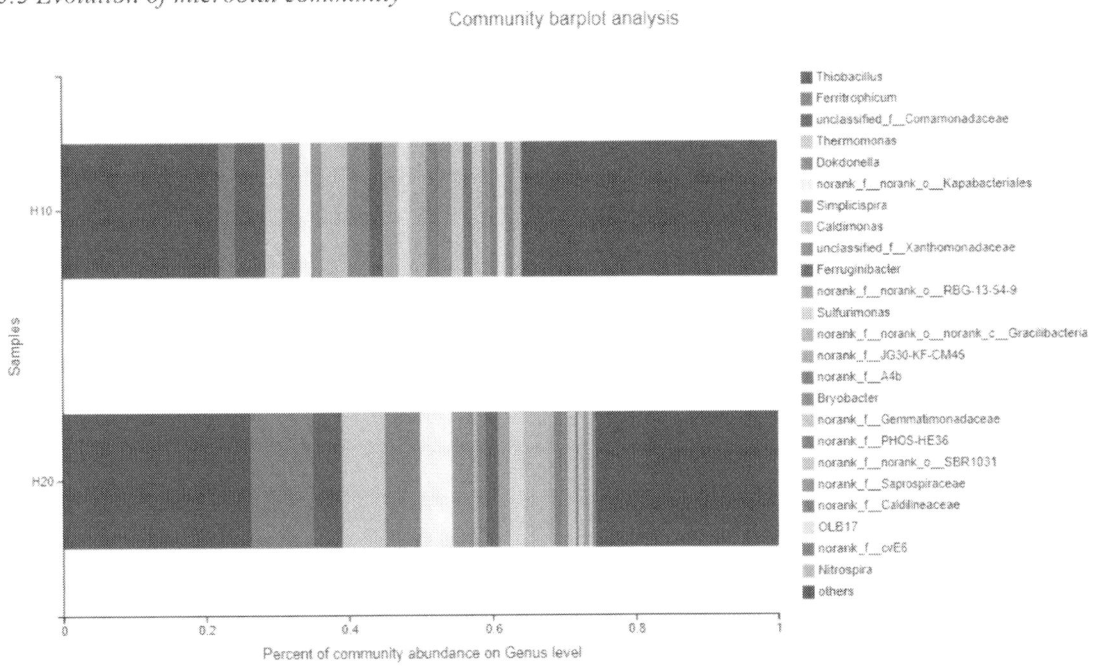

Figure 3. The distributions of bio-samples at the genus level in the reactors
(The samples H20 and H10 were collected from R_1 and R_2 reactors, respectively).

The taxonomic classification results of the microbial samples are shown in figure 3, in the two samples, the most dominant genus at the genus level was *Thiobacillus*, while *Thiobacillus* is typical denitrifying bacteria in the SAD process [10,11], its enrichment also shows that the SAD process would be a dominant process in the reactor. The higher relative abundance of *Ferritrophicum* in R_1 compared to R_2 may have contributed to the denitrification process, since *Ferritrophicum* was also shown to be denitrifying bacteria [12]. The appearance of *Dokdonella* in the reactor may be helpful for the reduction of sulfate productivity in the denitrification process, because it is heterotrophic denitrifying bacteria [6], and the relative abundance of *Dokdonella* in R_1 is higher, this may be related to the organic carbon source contained in the oyster shell powder in the F_1 filter.

4. Conclusion

According to the results of the above experiments, the following conclusions were drawn:

(1) For the SAD process, the use of oyster shell powder resulted in better denitrification performance and lower sulfate productivity than calcium carbonate powder.

(2) The method of intermittent vibration is more helpful for denitrification than backwashing.

In general, the R_1 reactor achieved better denitrification performance than R_2, this is due to the use of F_1 filter, and the F_1 filter may be more helpful to the enrichment of denitrifying bacteria in the reactor.

Acknowledgments

This work was supported by the National Key Research and Development Program of China (Grant No. 2020YFC1806402) and the Ningxia provincial Natural Science Foundation of China (Grant No. 2020AAC03272).

Appendix A. Supplementary data

Supplementary data associated with this article can be found, in the on online version

References

[1] Wang J L and Chu L B 2016 *Biotechnol. Adv* **34** 1103-12

[2] Liang B R *et al.* 2020 Investigation of different solid carbonate additives in elemental-sulfur-based autotrophic denitrification process coupled with anammox process *Environ. Technol. Innov* **20** 101149

[3] Sahinkaya E, Kilic A and Duygulu B 2014 *Water Res* **60** 210-7

[4] Chen F M, Li X, Gu C W, Huang Y and Yuan Y 2018 *Bioresour.Technol* **266** 211-9

[5] Liang B R, Zhang K, Liu D D, Yao S, Chen S T, Ma F, Wang Y Z and Zhu T 2021 Exploration and verification of the feasibility of sulfur-based autotrophic denitrification process coupled with vibration method in a modified anaerobic baffled reactor for wastewater treatment *Sci. Total Environ* **786** 147348

[6] Liang B R, Kang F, Wang Y, Zhang K, Wang Y Z, Yao S, Lyu Z N and Zhu T 2022 Denitrification performance of sulfur-based autotrophic denitrification and biomass-sulfur-based mixotrophic denitrification in solid-phase denitrifying reactors using novel composite filters *Sci. Total Environ* **823** 153826

[7] Liang J, Chen N, Tong S, Liu Y J and Feng C P 2018 *Chemosphere* **212** 954-963

[8] Sengupta S, Ergas S J and Lopez-Luna E 2007 *Water Environ. Res* **79** 2519-26

[9] Zhou W L, Sun Y J, Wu B T, Zhang Y, Huang M, Miyanaga T and Zhang Z J 2011 *J. Environ. Sci* **23(11)** 1761-69

[10] Xing W, Li J L, Li D S, Hu J C, Deng S H, Cui Y W and Yao H 2018 *Environ. Sci. Technol* **52(14)** 7867-75

[11] Vandekerckhove T G L, Kobayashi k, Janda J, Nevel S V and Vlaeminck S E 2018. *Bioresour.Technol* **257** 266-273

[12] Wan D J, Li Q, Liu Y D, Xiao S H and Wang H J 2019 Simultaneous reduction of perchlorate and nitrate in a combined heterotrophic-sulfur-autotrophic system: Secondary pollution control, pH balance and microbial community analysis *Water Res* **165** 115004

A Scalable and Reconfigurable Machine Learning and Deployment Platform for Power Plant

Chen Zhu[1,*,a], Xu Renbo[1,b], Han Xu[1,c] and Wang Jian[1,d]

[1]Dalian Power Plant of China Huaneng, China

[*,a]196534887@qq.com,[b]artheriar@163.com,[c]hxhndc2012@163.com, [d]549470028@qq.com

Abstract. With the rapid development of the Internet of Things(IoT) and industrial automation systems, the data accumulated by the industry is exponentially increasing. With the dynamics of the system, this platform should support different, possibly diverse types of models with different resource requirements. The life-long data collection and analysis for the complex coal-fired Power Plant requires a Machine Learning (ML) and deployment platform that is scalable and reconfigurable. This paper proposes a scalable and reconfigurable ML platform for a power plant based on docker technologies that support online model deployment, execution, and scheduling. In order to support the model retraining, a mechanism is proposed to manage the execution of the model training and the seamless transitions between models, without interrupting the online operation of model serving. This platform has been deployed in a power plant with two coal-fired units for about five months. Results of the field test prove that this system is flexible, model reconfigurable, and can achieve smooth model switches with minimal overhead.

1. Introduction

With the rapid development of the IoT and industrial automation systems, the data accumulated by the industry is exponentially increasing. Because the equipment data contains important information in the industrial process, how to extract valuable information from the massive industrial data has become the demand of industrial production enterprises[1]. In order to use a large number of industrial data to create economic value for enterprises, more and more researchers and engineers use machine learning to carry out data-driven modelling of industrial equipment. However, compared with traditional commercial big data modelling, industrial big data has completely different characteristics: industrial data mainly comes from sensor networks, which must be processed in a timely and uninterrupted manner; industrial big data acquisition and use has a strict location and scope limits to comply with enterprise-related data security standards; and also special, complex industries with a long life cycle[2]. In the process, the industrial model will evolve with the aging of equipment and the renewal of the process. However, the computing power and technical support of industrial large data platforms are relatively limited[3]. The usual design is to update the model independently of the large data platform and through manual completion, which makes the update cycle of the model long and high cost. The unique demand of industrial big data makes the practical use of those platforms face many difficulties.

Based on the prediction requirements of the main operating parameters of the thermal power unit, this article is carried out from the perspective of big data processing, modelling, and system demonstration, through streaming data processing, thermal power unit big data modeling, and data visualization display. The main work of the paper is as follows:

- An industrial big data modeling and service platform was built based on the Spark data flow engine and docker[4]. On the basis of existing data streams, historical data storage, real-time data management, and elastic scaling of resources, a continuous Machine learning modelling and management support.
- An innovative model updating scheduling algorithm based on model adaptability error is proposed. Because model updates are usually computational and resource-intensive, the limited resources of industrial computing platforms will be exhausted if the model updates are not properly scheduled. By evaluating the adaptability error of the model online, considering the evaluation results and the updating cost of the model comprehensively, the priority of model updating scheduling is determined, and the quality of service of the whole model of the system platform is improved.

The platform has been implemented on Rancher through Spark data stream engine and TensorFlow machine learning. The prototype system has been deployed on a 1000MW thermal power unit to model and predict eight key parameters of the system. In the process of system deployment, many models have been evolved, such as overhaul of units. Site testing and 10 months of continuous service show that the platform can adapt to the model evolution in industrial scenarios, and maintain the quality of service of the model at a high level when the system changes. At the same time, the cost of the system is very small, and it is very suitable for the use of large industrial data platforms with limited resources, especially the edge computing gateway.

2. RELATED WORK

With the advent of the era of big data, the large amount of data generated and accumulated in industrial processes has also accelerated the transformation of the industrial field. In order to enable industrial data to continuously and stably provide value to industrial systems, the platform construction and model services of industrial big data have become the focus of research.

Hyunseok Chang et al. [5] proposed a new cloud computing model called Edge Cloud, which solves the specific problems of edge computing by extending the cloud model based on traditional data centers and service nodes located at the edge of the network. Cun Ji et al. [6] proposed an industrial big data integration and analysis platform, which can well support offline storage and real-time computing of data. Fan-Hsun Tseng et al. [7] discussed the need for scalability of industrial big data platforms and proposed a fuzzy-based real-time auto-scaling mechanism for dynamic allocation of resources. B.I.Ismail et al. [8] evaluated the application of Docker containers in industrial big data, and their experimental results show that Docker provides rapid deployment, small footprint and good performance, which makes it possible to become a viable edge computing platform. support. In 2017, Google proposed a TensorFlow-based machine learning platform at the KDD conference, which supports offline training and online service of models. The platform has also been deployed on the Google Play app store, reducing installation time and increasing the app install rate by 2% [9,10]

3. PLATFORM ARCHITECTURE

In this section, the major architecture of this platform is introduced.

3.1. Major Structure

The overall technical process of the system is shown in Figure 1. Streaming data processing mainly involves related big data computing architecture and database technology. The processed data is used for model training and parameter prediction. Model construction mainly involves time series modeling algorithms in deep learning. The data is compared and displayed, and the key technologies involved are the Django framework and the Echart chart library.

Docker is a new lightweight virtual machine framework that has developed rapidly in recent years. Since the Docker engine is based on the LXC (Linux Container) technology, it ensures the repeated loading of the kernel when each container is executed, and at the same time, by sharing the kernel with the host, the application startup speed is accelerated and unnecessary memory consumption is reduced. Docker container technology ensures that each application can run independently with the host infrastructure, and can help developers deliver code, test and deploy faster, greatly shortening the entire deployment cycle.

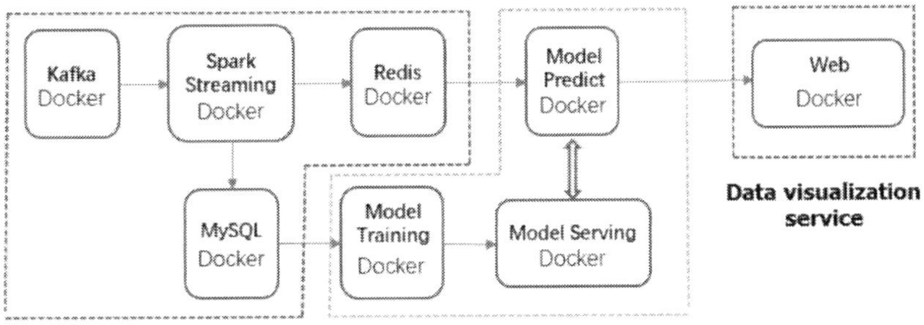

Figure 1. The major structure of our platform

Kafka is a message system based on publish and subscribe. It not only provides a distributed, partitioned, and persistent redundant backup log service, but also provides a unified, high-throughput, high-reliability, and low-latency platform for processing real-time data. Since its throughput can increase linearly with the expansion of the cluster, it is widely used in the field of big data collection and processing.

Spark Streaming is a real-time stream computing framework based on Spark Core. It has the characteristics of high throughput and strong fault tolerance, and supports data sources including Kafka, Flume, and TCP Sockets. After processing, the data can be stored in a highly reliable storage system. , can also be displayed in real time.

3.2. Stream-based processing

Because there will be a large number of data-driven models on the big data platform of thermal power units, and the establishment of the model requires the platform to support offline batch data computing. Based on the trained model, the model can only provide services after the real-time device data stream is processed into the model input format. Therefore, in addition to supporting the offline batch data training of the model, the thermal power unit big data platform also needs to process the streaming data in real time to provide online services for the model. This framework adopts a streaming processing framework based on Kafka and spark streaming which has the following characteristics:

(1) It can receive and process a series of continuously changing data.

(2) The event time and processing time of data can be paid attention to according to specific application scenarios.

(3) It has the characteristics of high throughput and low delay, and can meet the response calculation time of millisecond or microsecond.

(4) Fast: The data can be analysed before it reaches the disk.

3.3. Trainer

Trainer is an abstraction of a series of processes such as model training on the platform. It is mainly responsible for model training, in addition to data ETL, feature engineering and model preservation. ETL is responsible for extracting data from the historical database and performing data cleaning. The timing of data extraction can be based on the actual model update cycle or external triggers. The data is extracted using a sliding window, and the size of the window depends on the amount of training required by the model. Decide. It is this sliding window-based data extraction and preparation method that provides the basis for Trainer to automate model training. Figure 2 shows the Trainer's model automated training process.

IOP Conf. Series: Earth and Environmental Science 1044 (2022) 012004 doi:10.1088/1755-1315/1044/1/012004

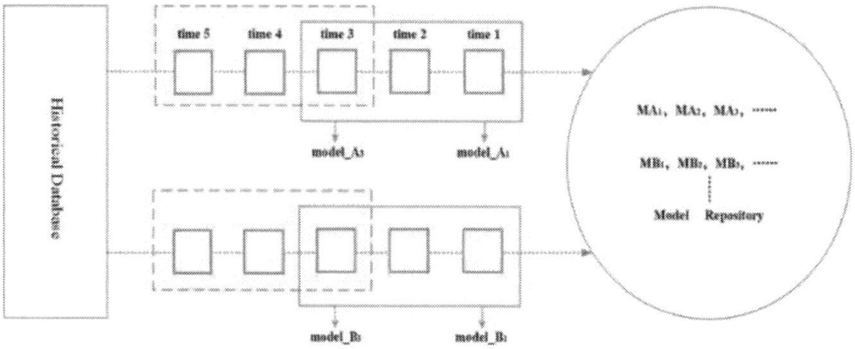

Figure 2. The dynamic training process

3.4. Scheduler

A platform that provides continuous services in resource-constrained scenarios needs to be able to ensure elastic scaling of platform resources. The resource elastic scaling provided by this platform mainly includes two aspects: 1. The resource management and control of the Docker container itself; 2. The resource elastic scaling based on Rancher.

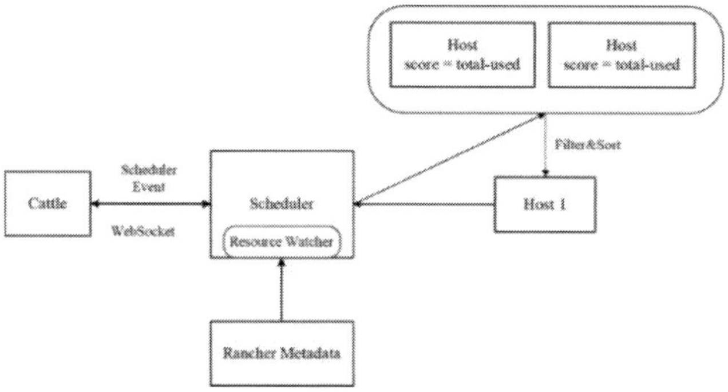

Figure 3. The major structure of the scheduler

In addition to the resource management and control capabilities of the Docker container itself, Rancher can also achieve scheduling by computing resources, which can further improve resource utilization for industrial big data platforms with limited resources. For scheduling by computing resources, Rancher mainly uses the scheduler service in the basic service for implementation. Its scheduler's computing resource allocation mechanism is shown in Figure 3:

During the scheduler service process, Rancher Events will be subscribed, that is, events related to the Scheduler. Once a scheduling request occurs, the Scheduler will send the calculated appropriate scheduling target to Cattle based on the Websocket method. Resource watcher will continuously calculate the usage of resources according to the data changes in Rancher metadata, and finally sort according to the remaining amount of resources as the scheduling target to achieve the goal of scheduling according to computing resources.

4. EXPERIMENTS

Our platform is used for a power plant with two 350MW coal-fired units with around 21142 physical sensors and 3G data per day. Here, models are built to forecast the future with different time steps, normally 15~60 seconds.

4.1. Basic facts

In order to verify the application effect of the model update scheduling algorithm based on the model adaptive error proposed in this paper, this section conducts multiple sets of tests on the performance of the adaptive error-based scheduling algorithm on the forecasting system and compares it with the regular update scheduling of the model and the training time based on the model. Two traditional model update scheduling algorithms for scheduling are compared.

The test uses three prediction targets with large fluctuations in thermal power units: power, active power, and main steam temperature. The establishment of the three prediction models uses the steady-state data of thermal power units when the load is 450MW, that is, the load is in the range of 430MW-470MW. We use the unsteady data of load between 450MW-1050MW as the actual test data. Every 100MW increase in load corresponds to 2h of test data. All the test data are divided into 6 intervals, with a total of 14,400 pieces of data for 12h. Table 1 shows the relevant information of the three prediction models.

Table 1. Basic facts for three models

Object	Error	Training Time	Update Interval
Power（PO）	0.23	55min	4h
Active Power（AP）	0.20	43min	4h
Main Steam Temp（MST）	0.12	34min	4h

4.2. Results Comparison

The training time of the comprehensive model and the scheduling strategy when different scheduling algorithms face multiple models, we set the update period of the three prediction models to the same 4h. Based on 12h of unsteady test data, the platform builds a real-time data stream every 3 seconds for the above three models. On this basis, we successively conduct experiments on two traditional model update scheduling algorithms and adaptive error-based scheduling algorithms. In order to easily show the error comparison of the three scheduling algorithms, for the traditional model update scheduling algorithm, we will also calculate the corresponding model adaptation error. The model adaptability error in the test is calculated every hour, and the calculated data is the prediction data of the model in the first 2 hours. In the test, we introduce the system adaptability error to represent the service quality of the three online models on the current simulation platform. The system adaptability error is obtained by the average value of the adaptability errors of the three models. Its trend chart and its system adaptability. The error data is as follows:

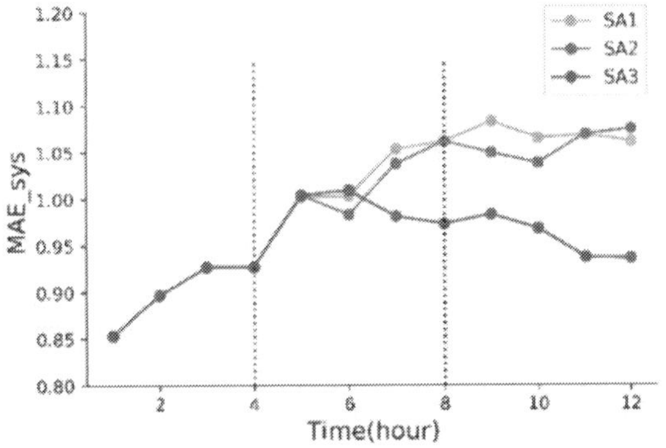

Figure 4. Trends of system adaptive error under three scheduling algorithms

SA1 adopts the model random scheduling method on the basis of periodic updates. Compared with calling SA2 with a short training time, it can be seen intuitively from Figure 4 that the two traditional scheduling algorithms SA1 and SA2 are in the whole time period. There is little difference in performance on systematic fitness errors. For SA3, in the period of 0-4h, since the weighted average MAE of the model calculated by the online evaluation in the past period did not exceed the allowable error of the model, if the online evaluation did not pass, the trend of the system adaptive error under SA3 was the same as that of SA1, SA2 is the same. But in other time periods, the performance of system adaptive error under SA3 is obviously lower than the other two scheduling algorithms. In the three-time periods (0-4h, 4-8h, 8-12h, SA2 has different system adaptability errors compared to SA1, and the average system adaptability error in the overall 12h Compared with SA1, SA2 is 0.7% lower than SA1, and the service quality of the model is slightly improved. However, the system adaptability error of SA3 in the three-time periods is not higher than that of SA1 and SA2, and the average adaptability error of the whole time period is compared with SA1. Compared with SA2, the reduction is about 5% and 4.3%. It can be seen that the scheduling algorithm based on the model adaptive error is better than the other two traditional model update scheduling algorithms in the performance of the overall model service quality.

Acknowledgment
This work is partially funded by the Huaneng Group Headquarters Technology Projects with No. HNKJ21-HF75.

References

[1] L. Da Xu, W. He, and S. Li, "Internet of things in industries: A survey," IEEE Transactions on industrial informatics, vol. 10, no. 4, pp. 2233-2243, 2014.
[2] M. Díaz, C. Martín, and B. Rubio, "State-of-the-art, challenges, and open issues in the integration of Internet of Things and Cloud Computing," Journal of Network & Computer Applications, vol. 67, no. C, pp. 99-117, 2016.
[3] F. Tao, Y. Cheng, L. Da Xu, L. Zhang, and B. H. Li, "CCIoT-CMfg: cloud computing and internet of things-based cloud manufacturing service system," IEEE Transactions on Industrial Informatics, vol. 10, no. 2, pp. 1435-1442, 2014.
[4] D. Merkel, "Docker: lightweight linux containers for consistent development and deployment," Linux Journal, vol. 2014, no. 239, p. 2, 2014.
[5] H. Chang, A. Hari, S. Mukherjee, and T. Lakshman, "Bringing the cloud to the edge," in Computer Communications Workshops (INFOCOM WKSHPS), 2014 IEEE Conference on, 2014, pp. 346-351: IEEE.
[6] C. Ji, S. Liu, C. Yang, L. Wu, and L. Pan, "IBDP: An industrial big data ingestion and analysis platform and case studies," in Identification, Information, and Knowledge in the Internet of Things (IIKI), 2015 International Conference on, 2015, pp. 223-228: IEEE.
[7] F.-H. Tseng, M.-S. Tsai, C.-W. Tseng, Y.-T. Yang, C.-C. Liu, and L.-D. Chou, "A Lightweight Auto-Scaling Mechanism for Fog Computing in Industrial Applications," IEEE Transactions on Industrial Informatics, 2018.
[8] B. I. Ismail et al., "Evaluation of docker as edge computing platform," in Open Systems (ICOS), 2015 IEEE Confernece on, 2015, pp. 130-135: IEEE.
[9] C. Olston et al., "TensorFlow-Serving: Flexible, High-Performance ML Serving," arXiv preprint arXiv:1712.06139, 2017.
[10] D. Baylor et al., "TFX : A TensorFlow-Based Production-Scale Machine Learning Platform," presented at the Proceedings of the 23rd ACM SIGKDD International Conference on Knowledge Discovery and Data Mining - KDD '17, 2017.

Energy Efficiency of Offshore Support Vessels

P J Ajith[1,a], T Bangar Raju[*1,b], Syed Aqib Jalil[1,c] and B K Chaturvedi[1,d]

[1] School of Business, University of Petroleum and Energy Studies, Dehradun. Uttarakhand State. India. Pin 248007

[*b]tbraju@ddn.upes.ac.in; [b]bangarraju@gmail.com; [a]pjajith@gmail.com; [c]aqibjalil@gmail.com; [d] bkchaturvedi@ddn.upes.ac.in

Abstract. The International Maritime Organization has brought forward the need to reduce CO_2 emissions from ships. There have been many deliberations in this regard; however, there is an absence of specific guidelines to monitor emissions in offshore support vessels. This study attempts to prioritize the various factors affecting the operational energy efficiency of these vessels. The objective of this study is to rank the factors and sub-factors based on feedback from various stakeholders. Three major factors namely Managerial, functional and Infrastructural and their respective sub factors were considered for the study. In total, seventeen sub-factors were taken for the study and feedback from twenty-two stakeholders were obtained. The Best worst method, which is an MCDM approach, is deployed to rank these factors based on feedback from various stakeholders. Infrastructure factors emerge as the leading factors among the three major factors: vessel performance monitoring, digitalized fleet management, and regulatory compliance. The study throws light for vessel operators to focus on infrastructure sub-factors for better energy efficiency in offshore vessels.

1. Introduction

The International Maritime Organization's Marine Environment Protection Committee [1] (MEPC 76) was held with a restricted program scheduled from June 10 to 17, 2021 on a virtual platform. It decided to put in place specialized operational control monitoring aspects to decrease CO_2 emissions in global shipping, which will be mandatory from 2023. The planned procedures are to incorporate the energy efficiency existing ship index (EEXI), more practical and improved objectives on the ship energy efficiency management plan (SEEMP), and the carbon intensity indicator (CII) rating with proper monitoring systems in place. It should be noted that the IMO was unable to release the final SEEMP guidelines, and the final CII rating methodology as the correction factors to these are yet to be determined. The above MEPC 76 has yet to address a specific methodology for ship energy efficiency management for vessels of 400–5000 GRT, consisting of a major world fleet of offshore support vessels. Hence, the only possibility is to follow [2] guidelines, where energy efficiency operational indicators (EEOI) can willingly be used for all existing and new ships. There are exceptional vessels that are not engaged in transport work, such as research vessels, tugboats, and offshore support vessels. In relation to the data collection system, currently, the IMO standard guideline for SEEMP [3] and amendments to came into force on March 1, 2016, shipping companies are in use. The Fourth IMO GHG Study 2020 [4] published mentions that the CO_2 emissions from offshore and service vessels comprise 9.4% of the total emissions from ships. In addition, these offshore and service vessels accounted for 28.4% of the global fleet of ships. The approach suggested by the IMO for the voluntary usage of the EEOI coefficient for offshore support vessels cannot be used because it is linked directly with the amount or quantity of work performed based on voyage coverage. Hence, from this gap to resolve the problem where a ranking of correct factors will help determine the top operational factors.

To rank the factors based on stakeholder responses in the shipping industry according to their prominence 37 expert responses were taken with well-experienced operational executives, owners, ship management, technical management, Port/Vessel operations, consultants, regulatory, etc. This study uses the best worst method (BWM) [5], a multi-criteria decision-making method, to identify the factors that can influence the energy efficiency of offshore support vessels. The study includes the managerial, functional, and infrastructure factors that impact the performance indication of the energy efficiency of such vessels. Section 2 reviews the relevant literature to establish the gaps in the segregation and grading of operational factors, especially due to the principal-agent approach. Section 3 describes the solution methodology and the steps taken into consideration during the survey. The resulting data and analysis using the BWM technique are presented in Section 4. It follows with a conclusion remark about the results, which identified the infrastructure factors as the most significant factors and their sub-factors for the monitoring of energy efficiency factors.

2. Literature Review

A large number of studies have been conducted on the energy and operational efficiency of ships over the past decade. Many researchers have studied the barriers to implementing energy efficiency and operational efficiency of ships in various shipping environments. Greenhouse gas emissions and energy consumption were investigated and correlated with the performance of river-going ships [6]. The navigational environment was found to be significant. Many other studies have investigated the environmental factors and their impacts on ship energy efficiency [7–11].Addressing the energy efficiency gap, Hannes Johnson and Karin Andersson conducted a couple of studies related to barriers to improving energy efficiency in shipping [12, 13].The principal-agent problem was also investigated in shipping from the study, and barriers to low-carbon shipping were studied [14].The principal-agent problem perspective was also recently studied to understand the influence of energy efficiency drivers and economic barriers [15].The effects of ship management models on ship operations toward better energy efficiency have been studied in [16].Navigational strategies have emerged as major factors in reducing exhaust emissions [17]. A novel energy performance index is proposed as, a better benchmarking tool for measuring the efficiency of cargo ships and bulk carriers [18]. Methods for increasing the energy efficiency of passenger ships through novel energy conservation measures are proposed [19]. Various technologies and their impacts are measured, and investment decisions are derived to improve energy efficiency. [20]. Bayesian belief networks for ship-port interface to measure ship CO_2 emissions and ship energy efficiency. Periodic hull cleaning and its impact on tanker ships' energy efficiency and fleet performance are discussed in [21]. The role of stakeholder involvement through operational measures for better energy efficiency in ships has been proposed in [22]. However, owners' interests and financial issues were found to be the main barriers for the same.

Research Gap: As per the formula for EEOI suggested by IMO, the mass of cargo carried is one of the parameters set. However, the intended usage of offshore support vessels is primarily not for cargo carriage, but to handle specialized operations in offshore activities. Hence, the factors that determine operations are variables based on priorities set by stakeholders. In this case, the results will not depend on the cargo carriage, but on the way each operation consumes fuel. The paper by Lützen, M[23] discusses the energy efficiency of working vessel structures, the main characteristics of working vessels such as offshore support vessels, the process of developing the framework with people involved, analyses of operations as a flowchart, and ends with a description of potential barriers(factors)to energy efficiency. However, the significance or ranking of these factors was not considered. Nam-kyun [24] attempted an efficiency evaluation index for ship operation energy by assigning weightage to various factors by an analytic hierarchy process between different shipping company operations. Operational KPIs focusing on principal-agent issues were not ranked.

In the paper Bartosz Glowacki & Cezary Behrendt [25], the authors present various analysed issues and shared conclusions related to the inappropriate results obtained by using a typical index for energy efficiency calculation for offshore support ships. As per the operational information, collected at an

AHTS type support vessel by considering the functional aspects of the vessel was also considered to evaluate the EEOI calculation by modifying the IMO formula. The author's findings indicate that the characteristics and arrangement of offshore support vessels (consisting of AHTS vessels analysed) contribute to energy efficiency assessment. The analysis concluded that the EEOI calculation should be derived for each type of ship as per their functional mode than a standard formula that is linked with the mass of cargo carried on board. This also implies the requirement to rate and rank different parameters for determining the EEOI for OSVs. Similar observations by Prill and Katarzyna [26] in the analysis to calculate the operational indicators concluded that criteria for performed transport work, which would be suitable for a particular vessel type, should be indicated in specific after applying it to a special research vessel. This finding indicates that in order to optimize the process of a specific ship in terms of its operational factors, needs to be carried out. Ship energy efficiency plan operational measures were prioritized using the fuzzy AHP model [27]. This brings the relative importance and strategic approach to selecting the most important measure that could be prioritized for better operational efficiency. However, there is a need to study and prioritize the various factors affecting offshore supply vessels. Based on the literature review, the factors that affect the energy efficiency of offshore support vessels were determined. The sub-factors were listed in Figure 1 and consolidated into three main factors: managerial, functional, and infrastructure.

3. Methods and Models

In the discussion above, the details of all the major factors and their corresponding sub-factors that affect the energy efficiency of offshore support vessel management are explained. We used the BWM method in this study to access and rank these factors and sub-factors. The BWM is a multi-criteria decision-making (MCDM) method applied to rank the alternatives by finding optimal weights for the criteria under study. In general, the BWM has some advantages over other similar MCDM techniques.

This requires less pairwise comparison data compared to other MCDM techniques. For example, in other MCDM techniques, such as the analytical hierarchy process, we compare each criterion with the others. In BWM, all criteria are compared with the chosen best and worst criteria by the decision-maker.

Results that are more consistent are achieved with BWM as it reduces the chance of errors due to the decision-maker's inefficacy in comparing the criteria.

Because of the advantages mentioned above, many authors have used BWM to solve real-life MCDM problems. [28] used this method to assess risks in business continuity management system. [29] applied this method to select suppliers based on their green innovation abilities. [30] accessed social sustainability factors for supply chains using BWM. [31] Identified an appropriate governance model for green port management using the BWM.

The stepwise procedure of BWM followed is as follows;

STEP 1: Identify a set of criteria (or factors).

This step identifies the set of criteria (or factors) that may be presented at multiple levels. The process of selection of criteria (factors and sub-factors) and their details are discussed.

STEP 2: Select the best and worst criteria.

In this step, we determine the best (B) (e.g., the most significant and most important) and the worst (W) (e.g., least significant, least important) criteria based on the preference data provided.

STEP 3: Determine the best-to-others vector:

This step determines the preference score (on a scale of 1 to 9) of the best criteria (B) over all other criteria. A preference score of 1 implies that the criteria are equally important to B, whereas a score of 9 implies that B is more important than those criteria. In this process, we obtain the best-to-others vector (A_B);

$A_B = (a_{B1}, a_{B2}, ..., a_{Bn})$, where $1, 2, ..., n$ the identified criteria. The vector elements are $a_{B1}, a_{B2}, ..., a_{Bn}$ represents the preference scores of criteria B over criteria $1, 2, ..., n$ respectively.

STEP 4: Determine the others-to-worst vector:

This step determines the preference score (on a scale of 1 to 9) of the other criteria over the worst criteria (W). A preference score of 1 implies that the criteria are equally important to W, whereas a score of 9 implies that the criteria are more important than W. In this process, we obtain the others-to-worst vector (A_w).

$A_w = (a_{1w}, a_{2w}, \dots a_{nw})T$. The vector elements represent the respective preference scores of criteria over criteria W.

STEP 5: Calculate the optimal weights:

In this step, the optimal weights are calculated by solving the linear programming model discussed in [5]. The objective function of the optimization model is to minimize the maximum absolute differences for all the criteria. The optimization model is solved as follows:

$$\begin{cases} Min \quad \underset{j}{Max} \left\{ \left| w_B - a_{Bj} w_j \right|, \left| w_j - a_{jW} w_W \right| \right\} \\ \text{Sub to;} \\ \sum_j w_j = 1 \\ w_j \geq 0, \ \forall j \end{cases} \quad (1)$$

The equivalent linear programming model for (1) is as follows;

$$\begin{cases} Min \quad \xi^L \\ \text{Sub to;} \\ \left| w_B - a_{Bj} w_j \right| \leq \xi^L \\ \left| w_j - a_{jW} w_W \right| \leq \xi^L \\ \sum_j w_j = 1 \\ w_j \geq 0, \ \forall j \end{cases} \quad (2)$$

By solving the linear programming model (2), we obtain the optimal weights $(w_1^*, w_2^*, \dots, w_n^*)$ and the optimal objective function value ξ^{L*}.

ξ^{L*} where represents the consistency index. Its value indicates the consistency of the pairwise comparisons provided by decision-makers. A value closer to zero indicates a high level of consistency.

The weights are calculated for decision-making scenarios with more than one level (having criteria and sub-criteria) for different levels following the BWM mentioned above. The weights of the second-level criteria can be multiplied by the weight of the corresponding first-level criteria to determine the global weights.

4. Results and discussions

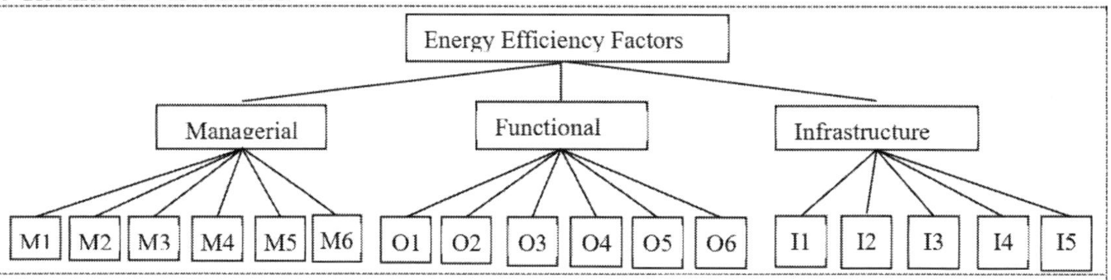

***Managerial*: M1,** Regulatory compliance[22][32]; **M2,** Capital constraints[22][32]; **M3,** current shipping practices[23][32]; **M4,** owner-charter issues[22][33]; **M5,** lack of incentives[32]; **M6,** change management issues[22][32].

***Functional:* O1:** Crew awareness and training [22][32][33], **O2:** Additional operational cost[22][32], **O3:** Berthing, maneuvering, passage, cargo operations[23][34], **O4:** Vessel age and operational service[22], **O5:** Trade and sailing area[22][35], **O6:** Turnaround time in ports[33].

Infrastructure: **I1,** Vessel performance monitoring [22][33][34]; **I2,** Technological complexities [28]; **I3,** Safety and reliability risks [22][32]; **I4,** Digitized fleet management [32]; **I5,** Operational mode monitoring [23][35].

Figure 1. Description of factors and sub-factors

Table 1. No of Experts Selected Energy Efficiency Factors as Best and Worst

Factors	Best	Worst
Managerial Factor	11	11
Functional Factor	5	19
Infrastructure Factor	21	7

Table 2. No of Experts Selected Sub-Factors of Managerial Factors as Best and Worst

Sub Factors	Best	Worst
Regulatory Compliance (M1)	13	2
Capital Constraints (M2)	8	3
Current shipping Practices (M3)	7	5
Owner-charter issues (M4)	6	4
Lack of incentives (M5)	3	6
Change management issues (M6)	0	17

Table 3. No of Experts Selected Sub-Factors of Functional Factors as Best and Worst

Sub Factors	Best	Worst
Crew awareness & training (O1)	16	2
Additional Operational cost (O2)	11	3
Berthing, Maneuvering, Passage, Cargo Operations (O3)	5	7
Vessel Age and operational service (O4)	1	7
Trade and sailing area (O5)	0	12
Turn-around time in ports (O6)	4	6

Table 4. No of Experts Selected Sub-Factors of Infrastructure Factors as Best and Worst

Sub Factors	Best	Worst
Vessel performance monitoring (I1)	13	2
Technological complexities (I2)	7	9
Safety & Reliability risks (I3)	5	9
Digitized fleet management (I4)	9	7
Operational mode monitoring (I5)	3	10

Table 5. Weights and Ranking of Energy Efficiency Factors

Factors	Optimal Weights	Rank

Managerial Factor	0.3155	2
Functional Factor	0.1924	3
Infrastructure Factor	0.4921	1

Table 6. Ranking of Sub-Factors of Managerial Factors

Sub Factors	Optimal Weights	Rank
Regulatory Compliance (M1)	0.3472	1
Capital Constraints (M2)	0.1861	2
Current shipping Practices (M3)	0.1657	3
Owner-charter issues (M4)	0.1634	4
Lack of incentives (M5)	0.0914	5
Change management issues (M6)	0.0462	6

Table 7. Ranking of Sub-Factors of Functional Factors

Sub Factors	Optimal Weights	Rank
Crew awareness & training (O1)	0.4324	1
Additional Operational cost (O2)	0.1951	2
Berthing, Manoeuvring, Passage, Cargo Operations (O3)	0.1107	4
Vessel Age and operational service (O4)	0.0964	5
Trade and sailing area (O5)	0.0216	6
Turn-around time in ports (O6)	0.1438	3

Table 8. Ranking of Sub-Factors of Infrastructure Factors

Sub Factors	Optimal Weights	Rank
Vessel performance monitoring (I1)	0.3921	1
Technological complexities (I2)	0.1316	3
Safety & Reliability risks (I3)	0.1234	4
Digitized fleet management (I4)	0.2517	2
Operational mode monitoring (I5)	0.1012	5

Table 9. Relative importance of factors and sub-factors

Factors	Factor Weights	Sub Factors	Local weights of sub-factors	Global weight of sub-factors	Global rank
Managerial Factor	0.3155	M1	0.3472	0.1095	**3**
		M2	0.1861	0.0587	**7**
		M3	0.1657	0.0523	**8**
		M4	0.1634	0.0516	**9**
		M5	0.0914	0.0288	**12**
		M6	0.0462	0.0146	**16**

Functional Factor	0.1924	O1	0.4324	0.0832	**4**
		O2	0.1951	0.0375	**11**
		O3	0.1107	0.0213	**14**
		O4	0.0964	0.0185	**15**
		O5	0.0216	0.0042	**17**
		O6	0.1438	0.0277	**13**
Infrastructure Factor	0.4921	I1	0.3921	0.1930	**1**
		I2	0.1316	0.0648	**5**
		I3	0.1234	0.0607	**6**
		I4	0.2517	0.1239	**2**
		I5	0.1012	0.0498	**10**

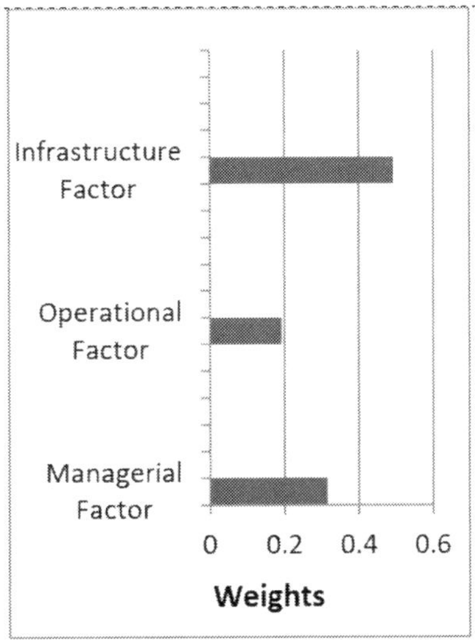

Figure 2. Relative importance of the main factors

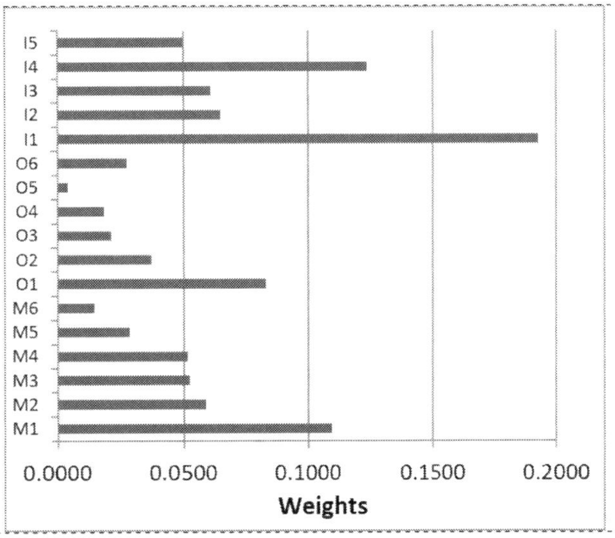

Figure 3. Relative importance of the sub factors

From the above, Figure 1 gives detailed description of factors and sub-factors which were shortlisted from literature review and based on feedback from the experts. Table 1 gives the detailed break up of feedback from experts for the energy efficiency factors. Infrastructure factors emerges as the leading factor among the three. Similarly, Table 2, 3 and 4 illustrate the feedback from experts for sub factors of Managerial, Functional and Infrastructure factors respectively. Table 5 illustrate the weights and ranking for three major factors. Infrastructure ranks first based on the weights derived. Table no 6, 7 and 8 showcase the ranking and weights for the sub- factors of Managerial, Functional and Infrastructural factors. As summarized in Table 9 as per the study, among the main factors, the infrastructure factor is ranked 1. It is considered as the most important factor with a global weighted value of 0.4921, which will impact the energy efficiency of offshore support vessels. This is followed by managerial factors (weight value: 0.3155) and functional factors (weight value: 0.1924) as shown in Figure 2. Among its sub-factors, vessel performance monitoring (I1) with weight (0.1930) showed the highest ranked value in the Global Rank out of the 17 sub-factors analysed. In contrast, the fifth criterion (O5) of functional factor, trade, and sailing area showed the lowest weight value (0.0042) in the Global Rank. This analysis with subfactor, namely vessel performance monitoring ranked as No.1, supports the findings of the recent 76th session of the IMO's Marine Environment Protection Committee, [1] which concludes that performance monitoring using advanced data capturing devices is already being implemented onboard the vessels. The sub factor ranked as the lowest trade and sailing area (O5), which is undoubtedly an important criterion, shows that, as per experts, this is a relevant factor common and applicable to all regions around the globe irrespective of demand drive in a specific country or operating region.

As per the analysis and the inference in Figure 3, global ranking weights 1, 2, 5, and 6 are for four out of five sub-factors: vessel performance (I1), digitalized fleet management (I4), technological complexities (I2), and safety and reliability risks (I3) are under the main factor infrastructure. This also concludes the relevance of the main factor infrastructure, which is due to the cumulative relevance suggested by experts rather than overrating a unique or specific subfactor. Hence, offshore support vessel owners have to focus on infrastructure development aspects that are more driven by the technology and monitoring of the same consisting of performance review, overcoming technological complexities, risk mitigation, digitization, and operational controls. The low ranking of operational mode monitoring in comparison with other sub-factors needs to be considered a natural consequence: there is no current control to review this activity very closely. Once other monitoring systems are in place, this could emerge as a key subfactor in a later stage.

Global weight rank 3 is for the sub factor for regulatory compliance (M1) classified under the main factor managerial factor. The recent IMO Maritime Environment Protection Committee's and European Union environmental regulations have enforced measures to reduce carbon emissions and stressed to intensify the monitoring need to reduce pollution and contribute to the international goal of net-zero carbon emissions by 2050.This is reflected in the expert opinion of key stakeholders of the survey as the member countries globally who had signed up with the IMO will seek to implement energy efficiency measures on offshore support vessel fleets.

The subfactor Crew awareness & training (O1) is at Global Weight Rank 4, which is classified under the main factor functional factor. There are IMO model training courses on energy efficiency measures for seafarers. However, the commitment and practical awareness of crew members' efforts and to investigate the implementation, adoption, and use of specific energy control measures and technologies onboard the offshore support vessel to improve energy efficiency is questioned by experts by the analysis findings. In addition, short-term crew contracts onboard the support vessels may be a reason for the crew not able to catch up with the commitment and awareness of energy efficiency measures implemented onboard. A possible solution is to ensure proper training and ensure that a mandatory monitoring system for crew efforts is in place on offshore support vessels. This will ensure practical concepts on implementation via the training of established skills and practices in place.

The major reason for the functional main factor having scored low weights and its sub-factors is that experts consider these factors to be less important in the current implementation, where the focus is to execute the major factors in place at the initial phase. However, this does not imply that the amount of energy savings from the functional factors is less, but gives a sense that these are not priority factors. Hence, it is important to consider all these factors together for the maximum outcome once the top-ranked ones are implemented.

5. Conclusions

Implementing effective energy-efficient measures, determining factors contributing to it, and developing a vessel-specific SEEMP is currently the aim of the offshore support vessel industry. The ultimate intention is the reduction in greenhouse gas emissions by the effective implementation of energy-efficient measures by identifying the correct factors contributing to the same. As reflected from the analysis, it is recommended to index the factors that will contribute to the energy efficiency of offshore support vessels based on a rating criterion and arrive at a method to determine the EEOI to the recommended set of factors for offshore support vessels. To determine the best energy-efficient enhancements, it is evident from the results that incorporating expert feedback from all decision-makers using multiple criteria decision-making techniques is the best methodology. The best worst method (BWM) was used to analyze the results as per the weight of each sub-factor at the global level and the major factor assessment level. Furthermore, it is proven from the results that analysis for evaluating key performance indicators is required, which shall be basic, straightforward, and reliable aspects that cover managerial, functional, and infrastructure factors. The proposed factor analysis with the BWM method for energy efficiency evaluation allows offshore support vessel companies to focus on and prioritize the energy efficiency focus per operation mode. Moreover, stakeholders and related establishments can utilize this method as a tool to evaluate their fleet of operations. The weighted index derived through the BWM multiple criteria decision-making technique developed in this analysis, along with supplementary studies, will be useful for developing a specific EEOI with the index score weighting to create a ship operation energy efficiency evaluation method for offshore support vessels.

References

[1] International Maritime Organization 2021 *Marine Environment Protection Committee MEPC 76),*https://www.imo.org/en/MediaCentre/MeetingSummaries/Pages/MEPC76meetingsummary.aspx (accessed 15th July, 2021)

[2] International Maritime Organization 2009 *GUIDELINES FOR VOLUNTARY USE OF THE SHIP ENERGY EFFICIENCY OPERATIONAL INDICATOR (EEOI): MEPC.1/Circ.684*

[3] THE MARINE ENVIRONMENT PROTECTION COMMITTEE 2016 *2016 GUIDELINES FOR THE DEVELOPMENT OF A SHIP ENERGY EFFICIENCY MANAGEMENT PLAN (SEEMP): RESOLUTION MEPC.282(70)* (MEPC 70/18*)*

[4] International Maritime Organization 2021 *4 Fourth IMO GHG Study 2020 (Safe, Secure and efficient shipping on clean oceans)* (london: Jasper Faber)

[5] Rezaei J 2015 Best-worst multi-criteria decision-making method *Omega* **53** 49–57

[6] Sun X, Yan X, Wu B and Song X 2013 Analysis of the operational energy efficiency for inland river ships *Transportation Research Part D: Transport and Environment* **22** 34–9

[7] Yuan Y, Li Z, Malekian R and Yan X 2017 Analysis of the operational ship energy efficiency considering navigation environmental impacts *Journal of Marine Engineering & Technology* **16** 150–9

[8] Ailong Fan, Xinping Yan, Richard Bucknall, Qizhi Yin, Sheng ji, Yuanchang Liu, Rui Song and Xiaping Chen 2018 A novel ship energy efficiency model considering random environmental parameters *Journal of Marine Engineering & Technology* **19** 1–14

[9] Wang K, Yan X, Yuan Y and Li F 2016 Real-time optimization of ship energy efficiency based on the prediction technology of working condition *Transportation Research Part D: Transport and Environment* **46** 81–93

[10] Wang K, Yan X, Yuan Y, Jiang X, Lin X and Negenborn R R 2018 Dynamic optimization of ship energy efficiency considering time-varying environmental factors *Transportation Research Part D: Transport and Environment* **62** 685–98

[11] Yan X, Wang K, Yuan Y, Jiang X and Negenborn R R 2018 Energy-efficient shipping: An application of big data analysis for optimizing engine speed of inland ships considering multiple environmental factors *Ocean Engineering* **169** 457–68

[12] Johnson H and Andersson K 2014 Barriers to energy efficiency in shipping *WMU J Marit Affairs* **15** 79–96

[13] Johnson H, Johansson M and Andersson K 2014 Barriers to improving energy efficiency in short sea shipping: an action research case study *Journal of Cleaner Production* **66** 317–27

[14] Rehmatulla N and Smith T 2015 Barriers to energy efficient and low carbon shipping *Ocean Engineering* **110** 102–12

[15] Longarela-Ares Á, Calvo-Silvosa A and Pérez-López J-B 2020 The Influence of Economic Barriers and Drivers on Energy Efficiency Investments in Maritime Shipping, from the Perspective of the Principal-Agent Problem *Sustainability* **12** 1–42

[16] Poulsen R T and Sornn-Friese H 2015 Achieving energy efficient ship operations under third party management: How do ship management models influence energy efficiency? *Research in Transportation Business & Management* **17** 41–52

[17] Perera L P and Mo B 2016 Emission control based energy efficiency measures in ship operations *Applied Ocean Research* **60** 29–46

[18] Zhang S, Li Y, Yuan H and Sun D 2019 An alternative benchmarking tool for operational energy efficiency of ships and its policy implications *Journal of Cleaner Production* **240** 1–11

[19] El Geneidy R, Otto K, Ahtila P, Kujala P, Sillanpää K and Mäki-Jouppila T 2018 Increasing energy efficiency in passenger ships by novel energy conservation measures *Journal of Marine Engineering & Technology* **17** 85–98

[20] Canbulat O, Aymelek M, Turan O and Boulougouris E 2019 An application of BBNs on the integrated energy efficiency of ship–port interface: a dry bulk shipping case *Maritime Policy & Management* **46** 845–65

[21] Adland R, Cariou P, Jia H and Wolff F-C 2018 The energy efficiency effects of periodic ship hull cleaning *Journal of Cleaner Production* **178** 1–13

[22] Dewan M H, Yaakob O and Suzana A 2018 Barriers for adoption of energy efficiency operational measures in shipping industry *WMU J Marit Affairs* **17** 169–93

[23] Lützen M, Mikkelsen L L, Jensen S and Rasmussen H B 2017 Energy efficiency of working vessels – A framework *Journal of Cleaner Production* **143** 90–9

[24] IM N, Choe B and Park C-H 2019 Developing and Applying a Ship Operation Energy Efficiency Evaluation Index Using SEEMP: a Case Study of South Korea *J. Marine. Sci. Appl.* **18** 185–94

[25] Bartosz GŁOWACKI and Cezary BEHRENDT (eds) 2017 *ENERGY EFFICIENCY OF OFFSHORE SUPPORT VESSEL The 58th International Conference of Machine Design Departments (Prague; Czech Republic, 6th to 8th September) (ICMD 2017)* (Prague: CZECH UNIVERSITY OF LIFE SCIENCES PRAGUE)

[26] Prill K and Igielski K 2018 Calculation of Operational Indicator EEOI for Ships Designed to Other Purpose Than Transport Based on A Research – Training Vessel *New Trends in Production Engineering* **1** 335–40

[27] Beşikçi E B, Kececi T, Arslan O and Turan O 2016 An application of fuzzy-AHP to ship operational energy efficiency measures *Ocean Engineering* **121** 392–402

[28] Torabi S A, Giahi R and Sahebjamnia N 2016 An enhanced risk assessment framework for business continuity management systems *Safety Science* **89** 201–18

[29] Gupta H and Barua M K 2017 Supplier selection among SMEs on the basis of their green innovation ability using BWM and fuzzy TOPSIS *Journal of Cleaner Production* **152** 242–58

[30] Badri Ahmadi H, Kusi-Sarpong S and Rezaei J 2017 Assessing the social sustainability of supply chains using Best Worst Method *Resources, Conservation and Recycling* **126** 99–106

[31] Munim Z H, Sornn-Friese H and Dushenko M 2020 Identifying the appropriate governance model for green port management: Applying Analytic Network Process and Best-Worst methods to ports in the Indian Ocean Rim *Journal of Cleaner Production* **268** 122156

[32] Michele Acciaro, Peter Nyegaard Hoffmann and Magnus Strandmyr Eide 2013 The Energy Efficiency Gap in Maritime Transport *Journal of Shipping and Ocean Engineering* **3** 1–10

[33] Nishatabbas Rehmatulla 2014 Market failures and barriers affecting energy efficient operations in shipping *Phd* University college london, UCL energy institute

[34] International Maritime Organization 2016 *STUDY ON THE OPTIMIZATION OF ENERGY CONSUMPTION AS PART OF IMPLEMENTATION OF A SHIP ENERGY EFFICIENCY MANAGEMENT PLAN (SEEMP) (Safe, Secure and efficient shipping on clean oceans)* (london: IMO)

[35] Styhre L 2010 *Capacity utilisation in short sea shipping (Doktorsavhandlingar vid Chalmers Tekniska Högskola* N.S., 3102*)* (Göteborg: Chalmers Univ. of Technology)

IOP Conf. Series: Earth and Environmental Science 1044 (2022) 012006 doi:10.1088/1755-1315/1044/1/012006

Overhead Transmission Line Anti-External Force Damage System

XueMin Leng [1,*,a], Jing Dai [2,b], Yang GAO [1,c], AoRan Xu [1,d] and Ge Jin [3,e]

[1] Shenyang Institute of Engineering, Shenyang, 110136, China;
[2] State Grid Liaoning Electric Power Co., Ltd., Shenyang, 110000, China;
[3] Shenyang Agricultural University, Shenyang, 110000, China

[*,a]lengxm@sie.edu.cn,[b]Bryan6001@163.com,
[c]gaoyangsie@163.com[d]xuaoran@163.com, [e]xz932871719@163.com

Abstract. In recent years, human production activities often lead to short-circuit grounding and other faults in overhead transmission lines, resulting in power supply interruption and economic losses, and traditional manual inspection methods are difficult to quickly find the cause and give early warning. In order to reduce the occurrence of such faults and ensure the reliability of power supply, the research on the anti-external force anti-damage system for overhead transmission lines is carried out in this paper. In this paper, the architecture of the anti-external force damage system for overhead transmission lines is designed, and an anti-external force damage algorithm for overhead transmission lines based on the improved R-FCN network model is proposed. have higher accuracy. The anti-external force damage system for overhead transmission lines designed and developed in this paper can realize real-time monitoring of the on-site environment of overhead transmission lines, and can quickly warn of various intrusions that may cause external damage, so as to avoid possible losses.

1. Introduction

In recent years, the huge power demand brought by the rapid development of economy has promoted the continuous increase of the grid structure of transmission network based on UHV transmission and other overhead transmission modes. In the protection area of power facilities, there are usually various large construction machinery and equipment to and from each other, which affects the safe and reliable operation of power facilities in the protection area of power facilities [1-3]. The geographical location and environmental conditions of overhead transmission lines are particularly complex, and the towers are widely distributed and large in number. The complexity of the environment of overhead transmission lines is bound to affect the reliability of the safe operation of transmission lines and equipment in special environments; Based on this, it is necessary to build a platform for preventing external force damage to the line corridor of overhead transmission lines so as to realize the online monitoring of overhead transmission lines. In References [4], a kind of transmission line anti-external force damage warning sign which can realize MPPT maximum light rate tracking is designed. The design of the device can effectively reduce the incidence of external force damage events of transmission lines. In References [5–8], based on 3G network, machine vision, image analysis and intelligent behavior analysis technology, the intelligent video surveillance system for preventing external force damage of transmission lines was studied. In Reference [9], a movable transmission line anti-external damage system is designed. The movable transmission line anti-external damage device can improve the safety of power grid and reduce the number of line power protection personnel. In Reference [10], an intelligent control platform for external force damage prevention of transmission lines based on multi-functional integration was designed, which can effectively control the external force damage of transmission lines. The external force damage of overhead transmission lines has the typical characteristics of randomness, dispersion and many types of external force damage. Relying solely on the traditional manual field monitoring means not only consumes a lot of manpower and material resources, but also has low monitoring efficiency, which cannot realize the multi-point and multi-type comprehensive monitoring of the external force damage of overhead transmission lines. Based on the study of advanced communication technology and online monitoring technology, this paper designs and develops the on-site monitoring and early warning system for the prevention of external force damage of overhead transmission lines, so as to find possible external force damage behaviors.

2. The overall structure and working principle of the anti-external force damage system for overhead transmission lines

As shown in figure 1 Design structure diagram of new anti-external force damage system. The structure of the intelligent early warning system for anti-external force damage of transmission lines is mainly composed of two parts: the hardware system of the front-end image video acquisition and the intelligent monitoring subsystem, and the backstage early warning software system. The working process of the whole system is as follows: the image video acquisition unit encodes and processes the video signal of the camera received by the video processing unit, and the intelligent monitoring subsystem analyses, processes and alarms the collected data in real time through data acquisition, storage, management and remote communication. The video information and alarm diagnosis information are transmitted to the monitoring center or the intelligent monitoring server through 3G network. After receiving the alarm diagnosis information and abnormal images, the monitoring center or the intelligent monitoring server will send the diagnosis information and abnormal images to the client in the form of SMS and MMS. The client can realize real-time tracking of the intrusion target through the background, and broadcast warnings by local voice.

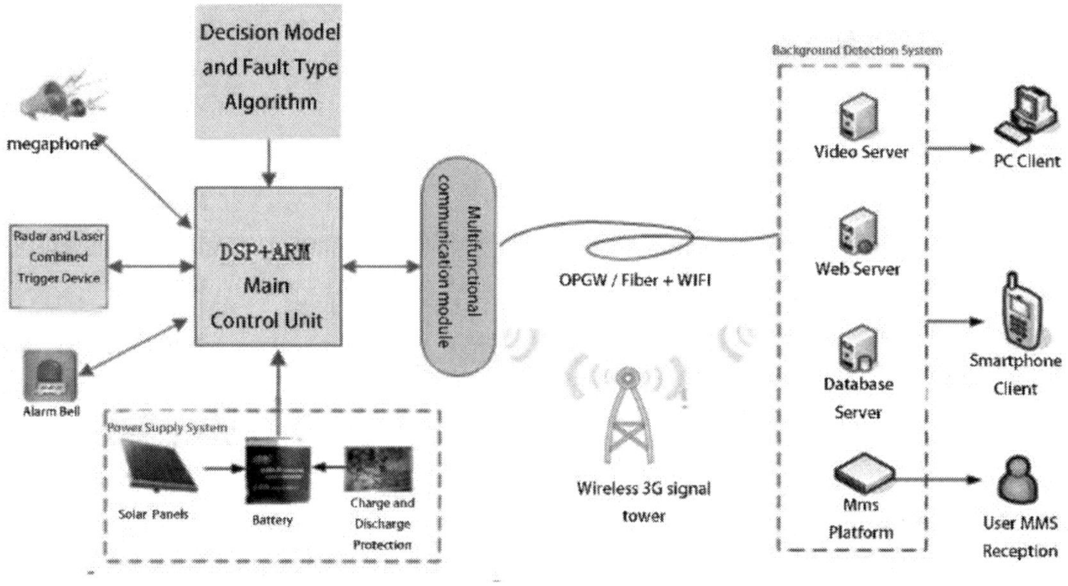

Figure 1. Design structure diagram of new external force damage prevention system

The front-end device architecture is mainly composed of the main control board, power supply system, radar monitoring device, three-dimensional acceleration sensor, temperature sensor, shortwave communication module, monitoring substation on the front tower and so on. The power supply of the front-end device is mainly supplied by the combination of inductive power generation and high-performance batteries. The real-time data acquisition of the front-end device on the field of overhead transmission lines is mainly obtained by radar ranging, 3D acceleration sensor inclination measurement and temperature sensor measurement. The real-time collected data are transmitted to the main control board of the front-end device. The type and distance of the intrusion are carried out through the decision model in the main control board, and the decision results are transmitted to the background system through low-power GPRS or 3G modules.

The background system completes its own data collection and display, stores the previous historical data, makes regular statistics, and gives the trend of change, so as to facilitate the operation of the dedicated staff to understand the change rule of the scene, but also considers the data interaction with other systems, plays a connecting link between the preceding and the following role, completes the function of transition middleware, can quickly understand through the PC client BS access mode, also can understand the latest data by SMS and APP through the mobile terminal. Software development considers scalability, practicality and accuracy, so that it can work independently and minimize the coupling with other systems, so as to facilitate installation, maintenance and upgrading.

3. Anti-external force damage algorithm of overhead transmission lines based on R-FCN model

The main function of the new on-line monitoring and judgment model for external force damage prevention of overhead transmission lines is to judge whether there are intrusions such as large-scale machinery approaching human theft of tower materials in the monitoring area through image data, monitoring whether there are kites and other foreign bodies hanging on the line ; and the classification and detection of large machinery and human targets are realized by video images, so as to predict whether there is external damage.

Target detection is often combined with deep learning to form a target detection model based on deep learning. The function graph of Region-based Fully Convolutional Networks (R-FCN) in the commonly used deep learning target monitoring model algorithm is easier to express, and it is easier to learn the scores and boundary boxes of classes. The principle of target detection using R-FCN network structure model is summarized as:

- Based on the feature extraction network ResNet, the feature extraction of the whole image is carried out, and the candidate regions are generated through the RPN network.
- Input the feature map output from the last convolution layer of the feature extraction network and the candidate regions recommended by RPN into the position sensitive score graph generation network.
- The position-sensitive fraction maps corresponding to each candidate region are input into the unified size of the position-sensitive RoI pooling layer, and then the border regression and classification of the candidate region are carried out.

R-FCN network structure model mainly includes: ResNet (Residual Networks); RPN network; Position sensitive fraction graph generation and position sensitive Ro pooling; Classification and regression. The R-FCN network structure model is shown in figure 2.

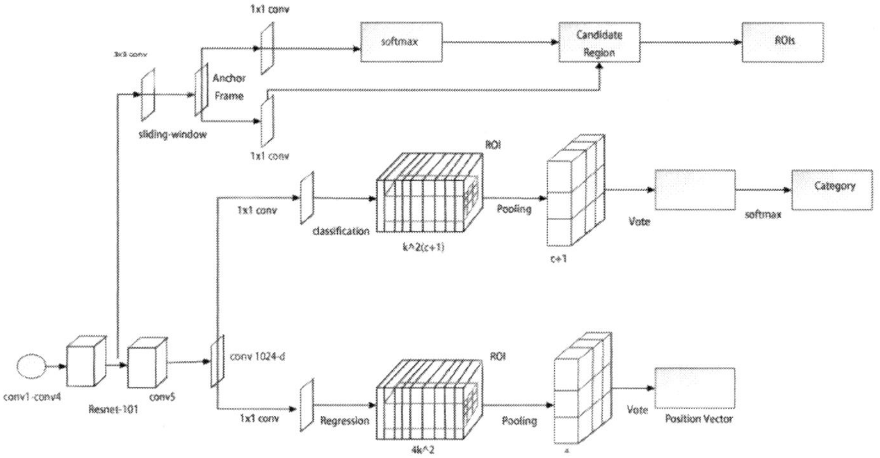

Figure 2. R-FCN network structure model

4. Verification of anti-external force damage algorithm for overhead transmission lines

4.1. Data set and experimental platform configuration of transmission line external force damage identification

Due to the limited total number of external damage images of transmission lines in the data set, the research group has accumulated monitoring images of transmission lines in cooperation with a power supply company project, and a total of 2400 images containing threatening objects have been obtained. Now, 2360 threatening images are divided into two types: external intrusion detection and hanging object anomaly detection and external damage detection. There are 856 images of external intrusion detection and hanging object anomaly detection (this paper uses image difference analysis algorithm to carry out anti-invasion and foreign object detection of overhead transmission lines), and 1544 images of external damage detection. The abnormal types corresponding to different overhead line pictures are shown in table 1.

Table 1. Abnormal types corresponding to different overhead line pictures

Line name	exception class	Line name	exception class
South Ring (dynamic risk)	Ground equipment	Shuang Bei Line(environment)	Crane
Qing Sha (dynamic risk)	Ground equipment	Hexiang (Construction Work)	Ground equipment
Rong Xing (construction risk)	Cranes and other types	Bian Bei (dynamic risk)	Ground equipment
Guang Hui (dynamic risk)	Crane	Bao Lai (dynamic risk)	Crane
Pan Shan (dynamic risk)	Crane	Liao You (dynamic risk)	Ground equipment
Liaohua (dynamic risk)	Ground equipment	Lu Yang (dynamic risk)	Ground equipment
Shan Bei	Crane	Liao He River	Ground equipment
Tianzhuangtai (environment)	Ground equipment		

The R-FCN network model used in this paper uses OpenCV to process images. All experiments are carried out on the Ubuntu system using TensorFlow deep learning framework.

4.2. Detection process of overhead transmission lines against external force damage based on improved R-FCN model

This paper uses transfer learning to solve. Based on the improved R-FCN model, the detection process of overhead transmission lines against external force damage is as follows:

- Using transfer learning algorithm to train data sets and generate pre-training network model.
- Extract the feature map of the whole image based on feature extraction network (feature extraction network using Resnet101 network).
- The candidate box recommendation network RPN uses the sliding window method to classify the anchor points of the eigenvalues extracted by the feature extraction network and regress the border to obtain the candidate box. The candidate box retains the candidate box with the highest score after the softening non-maximum suppression algorithm (threshold is set to 0.3, confidence is set to 0.1). The results of the background classification and the two branches of the reserved candidate region are aggregated to generate the candidate region into the position sensitive score graph generation network.
- The position-sensitive fraction maps corresponding to each candidate area are input into the unified size of the position-sensitive RoI pooling layer, and then the boundary regression and classification of the candidate area are carried out to realize the accurate identification of the external force damage of overhead transmission lines.

4.3. Experimental results analysis

In this paper, the anti-external force damage of transmission lines with abnormal types of ground cranes is taken as an example. The automatic identification and detection of external force damage types of transmission lines are carried out based on the data set. 544 pictures in the training set are trained and the training model after 50 iterations is saved. 1000 pictures are randomly selected for verification. The detection accuracy of the traditional R-FCN network model and the improved R-FCN network model is shown in figure 3. The accuracy of the improved R-FCN network model is 93.3%, while that of the traditional R-FCN network model is 91.5%, which fully proves that the improved R-FCN network model proposed in this paper can realize the accurate identification of the external force damage of transmission lines.

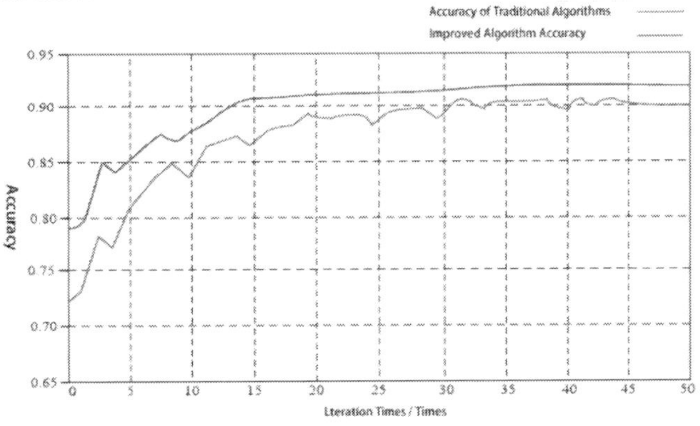

Figure 3. Detection accuracy of traditional R-FCN network model and improved R-FCN network model

The effect of automatic identification and detection of anti-external force failure types of transmission lines with abnormal types of cranes are shown in figure 4 and figure 5 respectively. The algorithm can accurately identify and detect the possible types of anti-external force failure on the site, so as to make early warning in time, facilitate the later tracking and remote monitoring of backstage personnel, and reduce the failure rate of overhead lines.

Figure 4. Inspection of tower crane and excavator under transmission line

Figure 5. Inspection for lower boom of transmission line

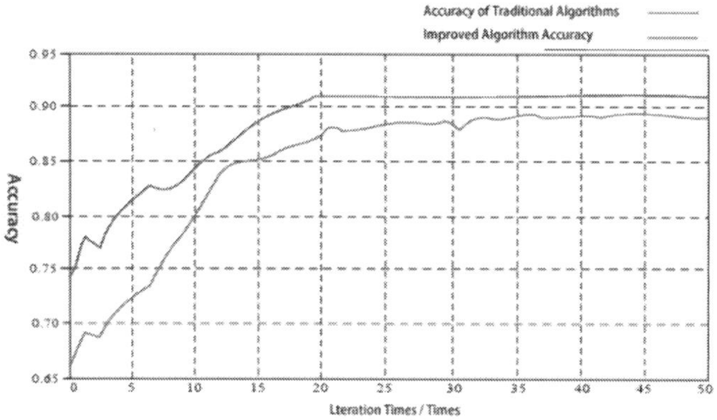

Figure 6. accuracy diagram of transmission line anti external force damage type of crane abnormal type based on traditional R-FCN network model and improved R-FCN network model

Based on the traditional R-FCN network model and the improved R-FCN network model, the accuracy of anti-external force failure type of transmission lines with crane abnormal types is shown in figure 6. The accuracy of anti-external force failure type of transmission lines with crane abnormal types of the improved R-FCN network model is 91.2%, and the accuracy of anti-external force failure type of transmission lines with crane abnormal types of the traditional R-FCN network model is 89.5 %.

5. Conclusions

In order to solve the problems existing in the process of traditional manual inspection of overhead transmission lines and realize the effective early warning of anti-external force damage of overhead transmission lines, this paper designs and develops a set of anti-external force damage system of overhead transmission lines, and proposes an anti-external force damage algorithm of overhead transmission lines based on improved R-FCN network model. Taking the anti-external force damage of transmission lines with abnormal ground cranes as an example, the automatic identification and detection of external force damage types of transmission lines are carried out based on the data set, and the pictures are randomly selected for verification. The detection accuracy of the traditional R-FCN network model and the improved R-FCN network model is compared and verified. The results show that the accuracy of the improved R-FCN network model is higher. The anti-external damage system developed in this paper can realize the real-time monitoring of the field environment of overhead transmission lines, and quickly warn various intrusions that may cause external damage, so as to avoid possible economic losses.

Acknowledgments
This work is supported by National Key R & D Program 08 Project 270 (Project No. 2020YFC0827003)

References

[1] Bolin Xie 2021 *Early warning of external damage of overhead transmission line anti-mechanical vehicles based on Internet of Things and positioning technology* (Fujian Institute of Engineering)

[2] Hai Yu 2019 Analysis of Prevention Measures for External Damage of Transmission Lines *J. Industrial Science and Technology Innovation* **1** 75-76

[3] JunYu Chen 2019 *Research on monitoring technology for external force protection of overhead transmission lines based on magnetic field measurement* (Southeast University)

[4] Atici Ada, Ouz Sunay M 2007 High Data-Rate Video Broadcasting Over 3G WirelessSystems *J. IEEE Transactions on Broadcasting* **53** 39-41

[5] Ye Zhang, Xinbo Huang and Xiaoxiong Chen 2015 Research on intelligent early warning system of transmission line against external force damage *J. High voltage apparatus* **51** 54-61

[6] Viola P and Jones M. Rapid 2003 *Proceedings of the 2001 IEEE Computer Society Conference on Computer Vision and Pattern Recognition* Object Detection Using A Boosted Aascade of Simple Features pp 511-518

[7] Viola P and Jones M J. 2004 Robust Real-Time Face Detection *J. International Journal of Computer Vision* **57** 137-154

[8] Freund Y, Schapire R E. 1996 *Proceedings of the ICMLl* Experiments with A New Boosting Algorithm pp 148-156

[9] Jinlong Liu 2017 Research and application of transmission line visualization and intelligent management system in Hengshui area (North China Electric Power University)

[10] Xiong Ge, Si Li and Jiangping Chu 2019 500 kV transmission line anti-break integrated management mode research *J. Mechanical and electrical information* **23** 16-19

Research on Fault Diagnosis of Switchgear Based on Temperature Cloud Image Technology

Zhiheng Chen[1,*,a], Ting Yuan[1,b], Mingfang Xu[1,c], Yang Yu[1,d], Zhen Xie[2,e] and Yuchen Liu[2,f]

[1]State Grid Panjin Electric Power Supply Company, Panjin, China
[2]Shenyang Institute of Engineering, Shenyang, China

[*,a]czh_pj@ln.sgcc.com.cn, [b]yt_pj@ln.sgcc.com.cn, [c]xmf_pj@ln.sgcc.com.cn, [d]yy_pj@ln.sgcc.com.cn, [e]xz932871719@163.com, [f]basinchen@outlook.com

Abstract. In the process of power system operation, contact aging, partial discharge and overcurrent faults are the main causes of switch cabinet faults. At the same time, it is difficult to install sensors inside the switch cabinet due to its closed operation. In this paper, in order to solve the over current problem of switchgear, according to the principle of high current heating, the thermal flow model of power switchgear is established, and the characteristics of heat flow inside the switchgear are studied. The infrared temperature detector is designed to collect the surface temperature cloud map of the switchgear. Combined with the internal heat flow law of the switchgear, the over current phenomenon in the switchgear is judged, so as to judge the operation state of the switchgear. In order to verify the correctness of the theory proposed in this paper, the operation state of the fault switch cabinet is designed for detection and comparative analysis. The test results show that the heat change caused by the over-current fault can be accurately fed back to the surface of the switch cabinet, which can be used as an effective means to diagnose the fault of the switch cabinet.

1. Introduction

The switch cabinet is one of the key equipment of the power system. The safe and stable operation of the switch cabinet is directly related to the power supply quality of the power system. The common thermal fault in the switchgear is the fault type with the most frequent occurrence. Many factors, such as the oxidation of contact and wiring end face, the decrease of contact surface caused by spring aging and the impact caused by sudden change of power load, will cause the internal overheating of the switchgear. If the switchgear is overheated for a long time, it will cause the aging of internal components and lead to the decline of insulation capacity, and eventually form a heat failure. In addition, the switch cabinet usually works in a strong electromagnetic environment, and the weak links in the switch cabinet are easily amplified, resulting in partial discharge and even fault arc. Fault arc will quickly heat the whole switch cabinet, easily lead to fault expansion and even the switch cabinet burned. With the increasing scale of power grid, how to accurately diagnose the internal temperature and fault types of switchgear is an important means to ensure the safe and stable operation of switchgear and avoid major safety accidents.Now infrared thermal imaging detection technology has become the most widely used detection technology after the five conventional non-destructive testing of ultrasonic (UT), radiography (RT), ET (ET), magnetic particle (MT) and penetrant testing. Currently, epidemic prevention and control and the power industry are the most important application

IOP Conf. Series: Earth and Environmental Science 1044 (2022) 012007 doi:10.1088/1755-1315/1044/1/012007

areas of infrared thermal imaging detection technology[1-2]. In this paper, by studying the physical law between the internal temperature of the switchgear and the temperature of the cabinet, the influence law of the internal fault of the switchgear on the temperature distribution of the cabinet, and combining with the artificial intelligence algorithm, an intelligent diagnosis method for the internal overheating fault of the switchgear is proposed.

2. Establishment of Heat Flow Model for Switchgear

In this paper, the metal armoured open switch cabinet is selected to establish a multi-physical field co-simulation model. Since the model of the switch cabinet is relatively complex and the establishment of the finite element software is relatively cumbersome, the model is established in SolidWorks software and imported into Fluent for analysis[3]. An important reason for the heating of switchgear is excessive load. When the switchgear load is too large, the current flowing through the switchgear A, B, C three-phase conductor increases, that is, the whole three-phase conductor is heating. In the simulation analysis, the uniform heat source is directly applied to the three-phase conductor to simulate. For the poor contact and fault arc of the internal components of the switchgear, there is a relatively obvious high temperature at a certain place. Based on the statistical analysis of the structural characteristics and fault probability of the switch cabinet, it is known that the most common locations of poor contact and fault arc in the internal components of the switch cabinet are mainly in the bus joint, circuit breaker joint and cable joint, and caused by excessive load. This is the main heating failure point of the switchgear shown in Figure 1, of which 1, 2 and 3 are bus A, B and C three-phase connections, 4, 5 and 6 are circuit breaker inlet A, B and C three-phase connections, 7, 8 and 9 are circuit breaker outlet A, B and C three-phase connections, 10, 11 and 12 are cable A, B and C three-phase connections.

In the simulation analysis, the heating failure of each fault point will be simulated, and the gas flow characteristics inside the switchgear, the internal temperature field distribution and the temperature distribution of the shell will be observed, which provides a reference for summarizing the general physical laws between the temperature distribution of the shell, the internal temperature distribution and the fault point[4]. Based on the characteristics of the switch cabinet heating fault, the probability of only one fault point in the switch cabinet producing fault is the largest, so only one fault point is simulated in the simulation. Model subdivision is a necessary step for simulation operation. The setting of model subdivision unit is too small, which will lead to a dramatic increase in the amount of calculation, occupy a large amount of resources and run for a long time. The model partition unit is set too large, and there may be no partition in the local area, resulting in inaccurate results, even errors, and unable to run. In the simulation process of the project, the height of each part of the switch cabinet is very different, reaching 2200 mm, but the shell thickness is only 3 mm. Therefore, in order to ensure the accuracy of the operation results, save computing resources and speed up the project process, the local segmentation method is used[5]. Set different partition restrictions for different parts, as shown in Figure 1. After several attempts of subdivision setting, it is determined that the upper line basis of the overall subdivision is 20 mm, and the shell and conductor copper strip are divided into two levels, which can achieve the purpose of accurate results and less resources.

Since the simulation model is a simplified structure of the real switch cabinet and the heat source is directly used to replace the current flowing through the resistance to generate heat, in the simulation analysis, the equivalent heat source ratio, diffusion coefficient and boundary thermal conductivity of the simulation should be designed and corrected based on the experimental data, which is divided into four steps.

The first step, heat source proportion correction. According to the heating power of the resistor wire in the test, the distribution of the three-phase conductor inside the switch cabinet is caused. In the simulation, when the temperature distribution law at the same key position of the conductor is reached, the ratio coefficient between the heat source power applied in the simulation and the heat source generated by the resistor wire is used.

The second step is diffusion coefficient correction. In the same way as the correction coefficient of the heat source, after the current is applied on the resistance wire, the corresponding heat source direction is given in the simulation, and the diffusion coefficient from the conductor to the air in the simulation is adjusted until the distribution difference of the key position of the internal temperature field of the switch cabinet is less than 5 %, and the diffusion coefficient from the conductor to the air is obtained.

The third step is the correction of boundary thermal conductivity. In the actual experiment, the switch cabinet is surrounded by air, and the boundary thermal conductivity of the switch cabinet shell is directly set in the simulation. The boundary thermal conductivity in the simulation is adjusted in the same way as that in the second step, and the distribution difference of the key position of the cabinet temperature field is less than 5 % under the same equivalent heat source, so as to determine the boundary thermal conductivity in the simulation.

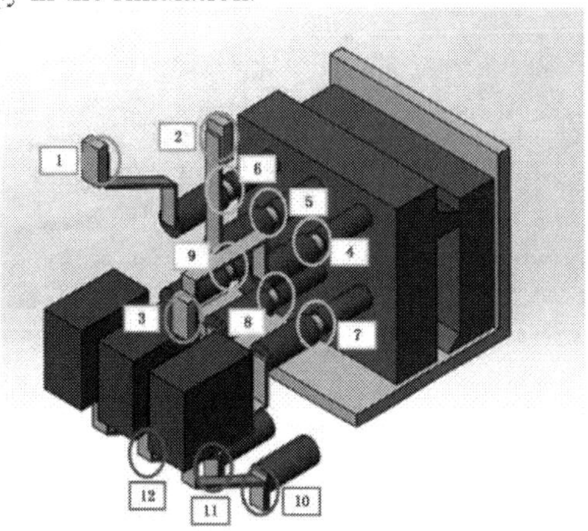

Figure 1. The switch cabinet

In the fourth step, the temperature distribution at the key position of the internal conductor surface, the temperature distribution at the key position of the internal air temperature field and the temperature distribution at the key position of the shell temperature field of the switchgear obtained from the overall comparison test are calculated, and the comprehensive difference is calculated. On the basis of the previous three steps, the proportion of equivalent heat source, the diffusion coefficient and the boundary thermal conductivity are fine-tuned, and the comprehensive difference is less than 3 %.

Using the correction method of the above parameters, the equivalent heat source ratio is determined as resistance wire heat source : simulation equivalent heat source = 1 : 0.84, diffusion coefficient 6.5W / m2, boundary thermal conductivity 5W / m2.

3. Heat Analysis of Cabinet Cable Contact Looseness Fault

The switch cabinet cable joint is similar to the circuit breaker joint, A phase and C phase are related to B phase, so the simulation analysis process can be simplified, only analyze A phase and B phase, C phase cable joint fault temperature field distribution is A phase mirror distribution.

Looseness of phase A joint

When the fault occurs at the A-phase joint, the power of the heat source is set to 300.06W to simulate the loose fault of the cable joint of the switchgear (357.21 * 0.84 = 300.06W). The gas velocity, the surface temperature distribution of the internal device, the internal temperature field distribution and the shell temperature distribution are obtained as shown in Figure 2, Figure 3, Figure 4 and Figure 5.

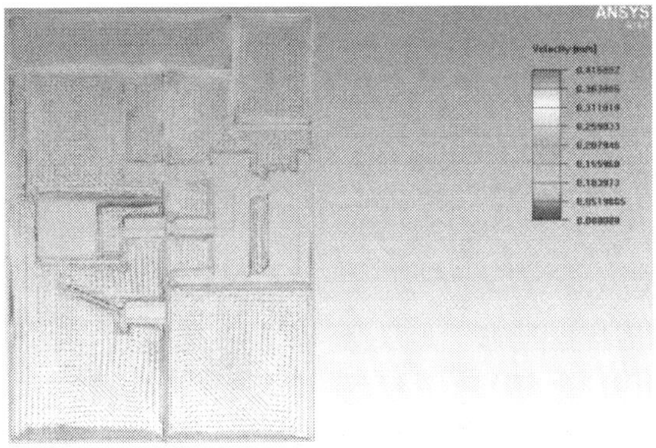

Figure 2. Gas velocity

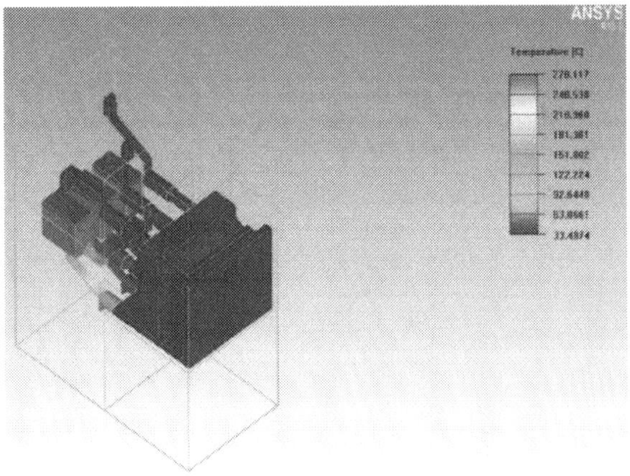

Figure 3. Surface temperature distribution of internal devices

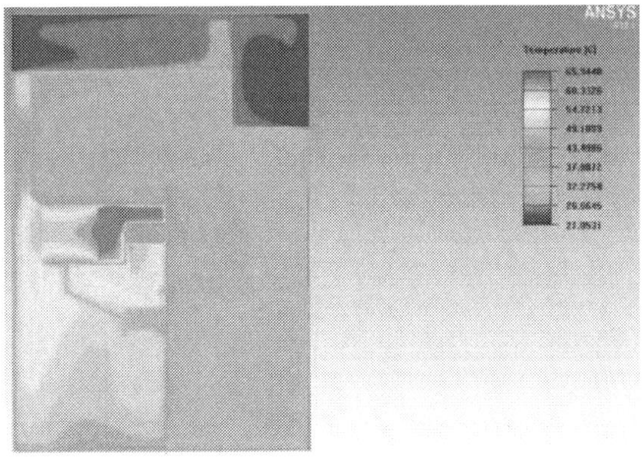

Figure 4. Internal temperature field distribution

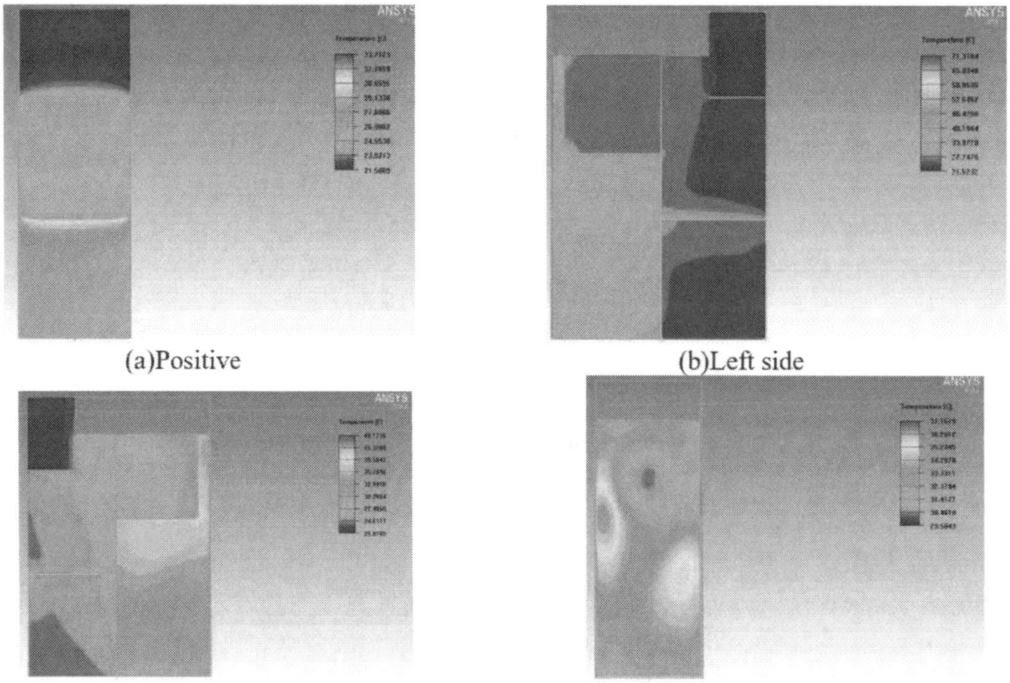

(a)Positive (b)Left side

(c)Right side (d)Back

Figure 5. Temperature distribution of shell

The simulation results of gas flow rate, internal device surface temperature distribution, internal temperature field distribution and shell temperature distribution obtained by different cable joint loosening thermal faults of circuit breaker were observed. For the internal gas flow rate of switchgear, the same rule can be found when comparing different cable joint loose heat fault conditions. With the increase of heat source power, the flow rate of gas is also enhanced, which leads to the accelerated thermal convection process of temperature. By observing the distribution of internal gas flow velocity, it can be seen that for the five mutually isolated chambers in the switch cabinet, the gas presents vortex phenomenon, and the flow velocity near the wall of each chamber is higher than that at the center. Among them, the highest flow rate is the cable room, especially the cable room close to the back cabinet body of the switch cabinet. Because the cable joint is located in the cable room, which is closest to the heat source and has the highest temperature, the heat of the cable room is mainly exchanged with the outside of the switch cabinet through the cabinet body, so the air flow velocity at the back cabinet body of the cable room is the largest.

By comparing and analyzing the surface temperature distribution of the internal components of the switchgear, it can be found that for different cable joint loosening heat faults, the temperature of the heat source point decreases gradually along the copper belt. The farther the distance from the heat source, the lower the temperature. For the internal temperature distribution of switchgear, according to the basic theory of thermal radiation, it is known that the farther away from the hot spot, the lower the temperature. It is not difficult to observe the temperature field distribution of different cable joint loosening heat faults. It is also found that the temperature decreases gradually from the heat source along the conductor to both sides, and then decreases from the conductor to the air. The area temperature of the cable room is the highest, because the heat source is located in the cable room. The highest temperature is located in the upper part of the cable room, because the air thermal expansion promotes its upward flow, so part of the heat is gathered in the cable room cabinet.

The temperature distribution of the switch cabinet shell was observed. The cable joint was symmetrically designed in the switch cabinet. The temperature fields of phase A fault and phase C fault must be mirrored. For phase B, the left side and the right side are almost symmetrical, so the difference in temperature distribution is also small, and only the shell temperature field distribution on the left side, the front and the back is compared. For phase A, the distance between the left side and the heat source and the conductor connected with the heat source is less than that of the right side, so only one side is analyzed in phase B. When the fault point is located in phase A, the temperature field distribution of the left side, the right side, the front and the back should be compared and analyzed at the same time. For the front temperature field distribution of the switchgear shell, the temperature distribution trend of the trolley room and the circuit breaker room is the same, the temperature of the upper part is higher than that of the lower part, and the temperature distribution of the trolley room is higher than that of the circuit breaker room as a whole. This is because the heat source is located in the cable room and directly transmitted to the trolley room, and the temperature at the junction of the circuit breaker room and the trolley room increases significantly. This is because the circuit breaker base is generally made of metal materials, and the thermal conductivity and heat absorption performance are better, which leads to the temperature rise here.For the temperature distribution on the left side and the right side, there is a certain corresponding relationship with the temperature distribution inside the switch cabinet, and the main temperature rise is located in the cable room. Compared with the left side and the right side of A phase, it can be found that the temperature of the left side is generally higher than that of the right side at the same position. The temperature distribution characteristics of the left side are closer to the internal temperature distribution of the switchgear, and the temperature distribution of the right side is more uniform. This is because the distance between the A phase and the left side is closer than that of the right side. The distribution on both sides of B phase is almost the same. The temperature distribution on the back of the switch cabinet shell shows the same law. The temperature distribution on the whole back presents a temperature distribution law that decreases from the top to the bottom. This is because the heat at the fault point is mainly transmitted to the inlet end, and the energy received at the outlet end is relatively small, so the temperature is low.

Based on the comparative analysis of gas flow rate, internal device surface temperature distribution, internal temperature field distribution and shell temperature distribution, it can be found that for different cable joints loose, the closer the contact is, the higher the temperature of the shell or conductor is. There is a certain corresponding relationship between the internal temperature distribution and the external temperature distribution of the switchgear. Due to the air thermal radiation and the heat conduction of the conductor in the heat transfer process, the temperature distribution characteristic phenomenon shown by the shell is lower than that of the internal, but the internal fault location can still be judged according to its characteristics.

4. Conclusion

In this paper, the heat generated by the switch cabinet fault and the heat flow mode inside the switch cabinet are studied. Through the numerical simulation and experimental measurement of the temperature field, the physical law between the temperature inside the switch cabinet and the temperature of the cabinet is summarized. On this basis, it is proposed to indirectly and accurately diagnose the internal fault type of the switch cabinet through the temperature of the cabinet, so as to further ensure the reliability of the operation of the switch cabinet and improve the quality of power supply of the grid. Taking the common cable contact loosening fault in switchgear as an example, the experimental analysis verifies the correctness of the proposed theory.

Acknowledgments

This work is supported by Science and Technology Project of SGCC (Research on Intelligent Prediction and Diagnosis Technology of Internal Fault Based on Big Data of Temperature Nephogram of High Voltage Switchgear) (5222PJ220003)

IOP Conf. Series: Earth and Environmental Science 1044 (2022) 012007 doi:10.1088/1755-1315/1044/1/012007

References

[1] Shan Jun. The application of infrared thermal imaging technology in the diagnosis of high-voltage electrical equipment. *Integrated Circuit Applications*, 2021, **38(01)**: 68-69.

[2] Jiamao Pang, Zhimin Yu and Xiangguang Chen. Research on thermal imaging fault detection system based on weibull distributed electrical system. *Journal of Physics*, 2021, **1941(1)**.

[3] Na Qu and Wen You, Design and fault diagnosis of DCS sintering furnace's temperature control system for edge computing. *PloS one*, 2021, **16(7)**.

[4] Liu Yinfeng, Zhang Jianping, and Li Aixiong. The application of infrared thermal imaging temperature measurement technology in metal machining. *China Metal Bulletin*, **2017(06)**: 70+69.

[5] Zhengwen Zhang et al. Application of edge computing in fault diagnosis of 10kV ring net switch cabinet. *Journal of Physics: Conference Series*, 2020, **1486(3)**.

Externality Assessment of Health Damage Caused by PM from PPV of China based on IPA

Yi Huang[1,a], Zongyuan Liu[1,b], Rongzhen Liang[1,c], Yanjie Jiang[1,d] and Xiaodi Ding[2,*,e]

[1]School of Management, Weifang Medical University, Weifang 261053, China
[2]School of Economics and Management, Weifang Institute of Technology, Weifang 250100, China

E-mail: [a]h20001119y@163.com; [b]lzyliuzyuan@163.com; [c]liangrongzhen123@163.com; [d]jiangyanjie_sdu@163.com; [*,e]xd.ding@wfit.edu.cn

Abstract. Health damage caused by pollutants is an important part of environmental externality. Its scientific and accurate evaluation is the basis of environmental policy-making and one of the recent research hotspots. In this paper, the Impact Pathway Approach (IPA) is used to evaluate the externality scale of health damage caused by Particulate Matter (PM) emitted by PPV (Private Passenger Vehicle) of China. Let the public, especially PPV owners, realize that the use of PPV has great health damage to others. Only by making active changes from themselves can reduce the negative externality. The core links of the IPA framework is the Concentration Response Function (CRF) and the determination of the average cost of each health damage observation point. In the end we got a total scale of health damage externality with the mean value of $2678522*10^4$ yuan, in which premature death took the largest part.

1. Introduction

The IPA used in this paper is a basic framework for accounting energy and environmental externalities. Its main function is to unify different externalities into a comparable unit, namely money (ExternE, 2005)[1]. For example, PPV consumes fuel during use, and its external effects include not only health damage caused by PM, but also greenhouse effect, acid rain, noise, congestion, etc. These different impacts can be compared and aggregated after being unified into currency in IPA.

The most two important links in the whole IPA analysis are: One is to determine the CRF according to pathological research and epidemiological investigation, which is also a dose response function (Dose-Response Function, DRF). It reflects the relationship between various health damage observation points invaded by air pollutants and the change of their concentration. The second link is to monetize the damage scale corresponding to the observation points by using certain environmental economic assessment methods. CRF is used in this paper, because the health damage caused by PM is caused by respiratory mechanism, which applies to concentration change and exposure mechanism.

CRF directly determines the object of the final monetization externality, and it is a pathological basis and core element for determining the infringement relationship. In case of health damage effect or disease, the patient will take economic actions to alleviate and treat the disease. These economic actions caused by health damage are called health damage observation points, such as consultation, drug purchase, hospitalization, and even death. The CRF is the quantitative relationship between the change of particle concentration and the observation points of health damage, which is generally obtained by

IOP Conf. Series: Earth and Environmental Science 1044 (2022) 012008 doi:10.1088/1755-1315/1044/1/012008

the investigation and statistics of health economists. Usually, the observation point of health damage must be observable first, so as to provide the premise for monetization.

2. Mathematical Model

2.1. Determination of CRF

There may be many forms of relationship between concentration and health damage. According to the research of Chuanjie Hong, et al. (2005) [2]and the fact that PPV source PM causes damage to human health through the atmosphere, this paper assumes that the form of CRF of PPV emission particles is the same as that of all particles in the whole atmosphere, which is:

$$HI_{PM} = HI_{BASE} \cdot e^{s \cdot \Delta C}, \Delta C = C_1 - C_0 \tag{1}$$

Among them, *HI* (Health Impact, HI) is the health damage observation point value. Before the change, it is HI_{BASE}, also known as the baseline value, attributed to C_0 before the change of PM concentration. It is obtained by multiplying the size of the exposed population which is Pop by the baseline incidence. HI_{PM} is the health damage observation point value of C_1 after the change of PM concentration ΔC.The unit of health damage observation point is the number of damage or occurrence cases per person per year. For example, if an individual has caught a cold twice a year due to the influence of particulate matter, the size of cold health damage observation point is 2. The s is the CRF coefficient, which represents the degree of health damage attributable to particulate matter and the sensitivity varying with concentration. The CRF coefficient is the key parameter in the CRF, which is mostly related to the statistical research of epidemiology. Sometimes, the RR (Relative Risk, RR) in equation (2), the ratio of the incidence of observation points before and after exposure, is also used to express the degree of health damage[3]:

$$RR = \left(\frac{HI_{PM}}{Pop}\right) / \left(\frac{HI_{BASE}}{Pop}\right) = e^{s \cdot \Delta C} \tag{2}$$

Then follows equation (3):

$$s = \frac{lnRR}{\Delta C} \tag{3}$$

2.2. Monetization evaluation formula

Assuming that the number of cases at each health damage observation point is S_i,the i represents different health damage observation points. Suppose that the different exposure rates adjusted by the population base in towns and rural areas are α, β, among which, $\alpha + \beta = 1, \alpha > \beta$. Then, for observation points, the number of cases in towns and rural areas are $S_i \cdot \alpha, S_i \cdot \beta$. It is assumed that the average cost of towns and rural areas at the observation points i are e_{ic} and e_{ir} respectively. Among them, $e_{ic} > e_{ir}$. Then the total externalities should be :

$$E = \sum_i \left(S_i \cdot \alpha \cdot e_{ic} + S_i \cdot \beta \cdot e_{ir}\right) = \sum_i [S_i \cdot \alpha \cdot (e_{ic} - e_{ir}) + S_i \cdot e_{ir}] \tag{4}$$

As equation (4) shows, since the average cost value of towns at each observation point is greater than the average cost value of rural areas, the whole externality increases strictly with the exposure rate level of towns.

3. Results Analysis

3.1. Calculation of CRF

First, take one of the health damage observation points, premature death, as an example to show the calculation process. According to the availability of various data, we take China's 2016 as the accounting year. First, the baseline of urban all-cause mortality was 567.37×10^{-5}, 617.6×10^{-5} in rural areas. According to the research of Jun Shen (2018)[4], the ratio of exposed population in urban areas is 96%

and that in rural areas is 13%. The calculated exposed population size is 76126.48 million and 7666.44 million respectively, accounting for 90.85% and 9.15% proportion respectively.

According to the reservation data of urban and rural PPV, we have obtained that the proportion of urban and rural PM emissions in the total amount of all sources is 2.081% and 0.6937% respectively. Based on the monthly PM data of 366 cities in China from January to December 2016, the national annual average value is 45.95 μg/m^3.

For different observation points and pollutants, the CRF coefficient "s" is different. It is generally believed that the observation points of chronic diseases are larger than those of acute diseases, and the fine particles are larger than other particles. The data of Peng Xie et al. (2009) [5]is used for the observation point of premature death, specifically when the concentration of PM increases by 10 μg / m^3, the incidence of premature death increases by 0.4%, and the relative risk RR is 1.0408. Next, according to the above data including exposed population, background value of PM concentration, baseline incidence, proportion of PPV emission and CRF coefficient, the number of premature deaths due to PM emitted by PPV in 2016 is calculated. There are 13184 cases in urban areas and 4395 cases in rural areas respectively. The 95% confidence interval is (6136, 20024), (1045, 6675).

According to the same ideas and treatment methods, combined with existing studies, other CRF coefficient of health damage caused by PM can be confirmed. As shown in Table 1:

Table 1. CRF coefficient at each health damage observation point caused by PM, baseline incidence and number of cases in 2016 of health damage observation points.

Health damage observation points		CRF coefficient	CRF 95% confidence interval		Baseline incidence	Number of cases: mean value	Number of cases: 95% confidence interval	
Premature death	urban	0.4[d]	0.19	0.62	0.5674[d]	13181	6136	20024
	rural				0.6176[d]	4395	1045	6675
Respiratory hospitalization		1.09[a]	0	2.21	0.5[b]	13388	0	28751
Cardiovascular hospitalization		0.68[a]	0.43	0.93	1.15[b]	18814	11748	26059
Acute bronchitis		7.9[b]	2.7	13	0.039[b]	10912	2811	24204
Chronic bronchitis		1.009[d]	0.27	1.30	0.694[d]	17130	4416	22401
Incidence rate of asthma		0.56[b]	0.27	0.84	5.61[b]	75126	35697	114297
Internal medicine clinic		0.49[b]	0.27	0.7	87.627[c]	1023154	557572	1477231
Pediatric Clinic		0.56[b]	0.2	0.9	216.298[c]	2896549	1015912	4735962

Data sources: a. Peng Xie et al. (2009)[5], b. Xiaoyun Liu et al. (2010)[6], c. 2017 China Health and family planning statistical yearbook (2017 Yearbook)[7], d. Haidong Kan et al. (2002)[8]. Note: since the CRF of some unimportant health damage observation points has not been established, it is not listed in Table 1 in this study.

3.2. Monetary assessment of health damage observation points

Monetary assessment is the last link of IPA analysis, it provides more general and comparative monetization information by monetizing the scale of health damage observation points obtained in the previous link. For example, the disease hospitalization observation points in Table 1 can be evaluated by the disease cost method in the revealed preference approach. In the 2017 Yearbook[7], the weighted average cost of hospitalization for respiratory system is 5476 yuan and for cardiovascular diseases is 21343 yuan.

Based on the average data (yuan/per case) of internal medicine clinics in 2015 of all provinces and municipalities directly under the central government adjusted by Jun Shen (2018)[4], this paper first

IOP Conf. Series: Earth and Environmental Science 1044 (2022) 012008 doi:10.1088/1755-1315/1044/1/012008

divides them according to regions, towns and rural areas, and adjusts the disposable income year-on-year to obtain the average data of outpatients according to regions, towns and rural areas in 2016. Then, using the population weighted average method, this paper obtains the average cost of medical outpatients in urbans in 2016 is 295 yuan/per case, and in rural areas is 89 yuan/per case. Like Xiaoyun Liu et al. (2010)[6], Desheng Huang et al. (2013) [9]and Jun Shen (2018)[4], pediatric clinic also uses this data.

Chronic bronchitis, acute bronchitis and asthma are obtained in the same way as internal medicine clinics. After obtaining the average data of three observation points by region, urban and rural in 2016, the population weighted average method can be used to obtain the average cost of three observation points by urban and rural in 2016, as shown in Table 2.

Table 2. Average cost of health damage observation points of chronic bronchitis, acute bronchitis and asthma in urban and rural areas in 2016

Urban: yuan/per case			Rural: yuan/per case		
Chronic bronchitis	Acute bronchitis	Asthma	Chronic bronchitis	Acute bronchitis	Asthma
5813	718	1663	1743	275	1444

Premature death observation points need to use WTP (Willingness To Pay) method to obtain statistical life value (VSL). This paper is based on Jun Shen (2018)[4] 2015 data and the ratio of per capita GDP of each region between 2016 and 2015. Then adjust the national VSL by region and obtain the national average VSL value 1422616 yuan/per case, according to the weighted average method.

Finally, the monetary assessment of the externality of each observation point of health damage caused by PPV PM is obtained by multiplying the number of cases in 2016 in Table 1 and the average cost. Therefore, based on some important setting conditions: urban and rural exposure rate (96%, 13%), the national average concentration of PM in 2016 was 45.95 μg/m^3, PM threshold 10 μg/m^3, the average cost of premature death VSL, respiratory and cardiovascular hospitalization do not distinguish between urban and rural areas, the monetary assessment of health damage externality caused by PPV PM in 2016 is shown in Table 3.

Table 3. Monetary assessment of health damage caused by PPV PM /10^4 yuan

Health damage observation point		mean value	95% confidence interval		Proportion of average value /%
Premature death		2500390	1021581	3798242	93.35
Respiratory hospitalization		7331	0	15744	0.2737
Cardiovascular hospitalization		40155	25074	55618	1.499
Chronic bronchitis	urban	9047	2332	11830	0.3378
	rural	273	70	357	0.0102
Acute bronchitis	urban	712	183	1579	0.0266
	rural	27	7	61	0.001
Asthma	urban	11350	5393	17269	0.4237
	rural	992	472	1510	0.037
Internal medicine clinic	urban	27422	14943	39591	1.0238
	rural	833	454	1203	0.0311
Pediatric Clinic	urban	77630	27227	126928	2.8982
	rural	2359	827	3856	0.0881
Total		2678522	1098564	4073789	100

It shows that the monetization externality assessment of health damage caused by PPV PM is small from the perspective of PPV source alone, accounting for only 0.036% of GDP in that year. Since PPV source account for 2.7747% of all PM sources, if converted, the externality of PM health damage will account for 1.3% of the whole GDP. Premature death accounts for a large proportion in the whole externality assessment. Although the number of cases is small, it absolutely is the largest externality assessment in all observation points of health damage due to the largest mean value of damage (VSL). In second place is the pediatric clinic, this paper found that the incidence calculated from the number of pediatric outpatients counted in the 2017 Yearbook[7] is much higher than that of other relevant studies, which reflects the sensitivity of PM to groups younger than 14 years old. In addition, with the increase of people's income and the liberalization of the "two child" policy, this part of externality must also be paid attention to.

4. Conclusion

Under the framework of IPA, the externality scale of health damage caused by PM emission during PPV use in China was evaluated. The two most important links are: The concentration response function, which lays a foundation for evaluating the externality monetization scale of health damage observation points; The average cost of each health damage observation point, which is determined under some important assumptions and by using appropriate externality monetary method, being used to calculate the monetary scale of health damage externality by using the CRF.

In terms of each health damage observation point, premature death is the largest part, with the amount of $2500390*10^4$ yuan mean value, accounting for 93.35% of all, indicating that the control of PPV particulate matter externality needs to focus on the system design of this part. Different from the small number of cases of premature death, the number of cases in pediatric outpatient service is very large, and it is more reasonable to directly compensate the person who occurs here.

Acknowledgments

The authors sincerely acknowledge the assistance from colleagues in WFMU and WFIT. And this paper is under the financial support from Shandong province Natural Science Foundation of China, with the project-The research on public value accounting of green travel and innovation incentive mechanism.

References

[1] ExternE. Externalities of Energy: methodology 2005 update[J]. Luxembourg: Office for Official Publications of the European Communities, 2005
[2] Chuanjie Hong, Haidong Kan, Bingheng Chen. The method of evaluating the health risk of urban air pollution: the fifth lecture on the quantitative assessment of the health hazards of air pollution to urban residents (Continued 5) [J]. Journal of Environment and Health, 2005, 22(1): 62-64
[3] Andrade C. Understanding relative risk, odds ratio, and related terms: as simple as it can get. J Clin Psychiatry. 2015 Jul;76(7): e857-61.
[4] Jun Shen. Research on the impact of $PM_{2.5}$ pollution on public health and social economy [D]. China University of Geosciences, 2018
[5] Peng Xie, Xiaoyun Liu, Zhaorong Liu, et al. Study on the exposure-response relationship of atmospheric particulate matter pollution among Chinese population [J]. China Environmental Science, 2009, 29(10): 1034-1040
[6] Xiaoyun Liu, Peng Xie, Zhaorong Liu, et al. Economic Loss Evaluation of Acute Health Effects of Inhalable Particulate Matter Pollution in the Pearl River Delta[J]. Journal of Peking University: Natural Science Edition, 2010 (5): 829-834
[7] National Health and Family Planning Commission. 2017 China Health and Family Planning Statistical Yearbook [M]. Beijing, China Union Medical College Press, 2018
[8] Haidong Kan, Bingheng Chen. The relationship between atmospheric particulate matter exposure and population health effects in China[J]. Journal of Environment and Health, 2002, 19(6): 422-424
[9] Desheng Huang, Shiqiu Zhang. Assessment of Health Benefits of Controlling PM2.5 Pollution in Beijing-Tianjin-Hebei Region [J]. China Environmental Science, 2013, 33(1): 166-174

Changes in Water Surface Boundary of East Juyan Lake in Northwest China Based on Remote Sensing Data

Lingqi Li[1, *, a], Jinqiang Lu[1, 2, b], Kai Wu[1, c], Bo Qu[1, d] and Bingbing Liu[1, e]

[1]Yellow River Institute of Hydraulic Research, Zhengzhou, 450003, China;
[2]School of Water Conservancy Engineering, Zhengzhou University, Zhengzhou 450001, China

Email: [*, a]lqli@whu.edu.cn; [b]ljq308227002@163.com; [c]wuukaii@foxmail.com
[d]qubo_edu_hohai@163.com; [e]liubingszy1231@163.com

Abstract. Lakes in arid inland areas are important indicators for reflecting the regional ecosystem security under climate change and human-related impacts. Understanding the evolution characteristics of lakes is helpful for eco-environment protection and management. This study applied the Landsat remote sensing data from 2002 to 2017 to analyze the water surface area changes of East Juyan Lake, a closed lake in northwest China. The results showed that the upward trends existed from 2002 to 2006 and were more significant from 2014 to 2017. The upward trends became gentle from 2007 to 2013. Regarding the seasonal characteristics, the water surface area in winter was almost the largest in the whole year, with an annual average of 51km^2, followed by that in autumn (50.45km^2). The annual average value in spring (48.16km^2) was larger than that in summer (41km^2). For the spatial changes, the lake boundary generally expanded from 2002 to 2009, and its eastern and western boundaries changed obviously after 2006. After 2010, the changes in lake boundaries tended to be gentle.

1. Introduction

Lakes are key parts of natural ecosystems. The shrinkage and expansion of lakes can indicate regional climate change and human-related impacts. Lakes in arid areas are more sensitive under changing environments and have important links to regional water cycles [1, 2]. Study on the evolution of lake boundaries in arid areas would be helpful to evaluate the impact on regional ecological environment change and promote the harmonious nature-human relationship [3].

The East Juyan Lake is the terminal lake of the Heihe River (the second largest inland river in China). It plays a vital role in blocking the progress of desertification. Because of the improper water resources allocation in the middle reaches, the East Juyan Lake began to largely shrink in the 1960s and even completely dried up. The ecological environment of East Juyan Lake wetland also significantly deteriorated [4]. Such an eco-environmental crisis in East Juyan Lake has attracted extensive attention from the nation and society. The government has decided to implement a unified dispatching and management project of the Heihe River water resources, to ensure the water demand of East Juyan Lake wetland and gradually restore the wetland functions for ecological ecosystems [5, 6]. Because of the water diversion for many years, the ecosystem restoration of the downstream Heihe River basin has improved significantly. However, in the context of global changing climate, water bodies of terminal lakes still face the risk of drying up [7, 8]. Therefore, research on the temporal changes and spatial distribution of the East Juyan Lake surface area has important practical significance for risk warning, water resources management of lakes, and social-economic-ecological development in the Lower Heihe River.

The research on the dynamic change of lake water areas and its influencing factors has always attracted the attention of scholars worldwide. In the early stage, most scholars attempted to use the historical geomorphological method of combining historical documents and geomorphic sediment survey data to analyze and restore the laws of lake evolution [9, 10]. However, this method could be easily constrained by the natural conditions or the recording methods and contents of historical data. With the increasing development of remote sensing technology, characteristics of large-scale, periodic, and real-time monitoring of remote sensing data have obvious advantages over traditional historical geomorphology methods [11, 12].

For example, Cao et al. [13] applied Landsat satellite data to the NDWI water body index and OSTU threshold method. They obtained the regional and temporal data sets of study areas of Hongjiannao Lake and further studied the effects of natural and human-related factors on varying lake water bodies. Zhang et al. [14] used Landsat 4, 5, 7, and 8 to analyze the dynamic changes of the Qinghai Lake area in Tibet Plateau. Emami et al. [15] used environmental and remote sensing image data to simulate the water surface change of Urmia Lake. The area of lakes can be mainly monitored by the imagery from remote-sensing technology. Both the moderate resolution imaging spectroradiometer (MODIS) [16] and Landsat images have been commonly used. Wang et al. [17] found that for small lakes, the Landsat satellite had higher accuracy in extracting water surface areas. Compared with MODIS data products, Landsat remote sensing data may provide more appropriate spatial and temporal resolutions and thus have been often employed in the case of long-term monitoring of small lake areas. Yue et al. [18] established a long-term data set of China's Great Lakes area by combining Landsat and hydrological data. Zhang et al. [19] used satellite imagery to study lakes with areas of more than one square kilometer in China and analyzed the reasons for lake area changes. Nitze et al. [20] proposed a method combining trend classification of multi-spectral Landsat indices and object-oriented lake detection, which helped to compare lake dynamics in different permafrost zones. Their results demonstrated the relationship between lake dynamics and variabilities of permafrost conditions, landscape location, and geology. This new method was found to be effective in various lake change analyses (transferability, scalability, and consistency) for the study area. Babaei et al. [21] extracted more accurate lake areas from Landsat 8 images through spectral indices and Otsu's threshold method. The used indices produced a high accuracy in distinguishing the water-land area boundaries.

In this paper, the lake boundary data were extracted using Landsat remote sensing images from 2002 to 2017, to analyze the temporal changes and spatial distribution characteristics for the water surface areas of East Juyan Lake after the implementation of water transfer in the Heihe River. This would be expected to provide a reference for lake water resources management and eco-environmental conservation in arid inland areas for the future.

2. Materials and methods

2.1 Study area

The lower sections of Heihe River are mainly divided into the East and West river reaches and flow into terminal lakes, which are named the East and West Juyan lakes, respectively. The study area was located in the East Juyan Lake, with longitude and latitude of 42°10'-42°20' N and 101°12'-101°19' E (Figure 1). Controlled by Mongolian air mass all year round, it belongs to a temperate continental climate, with sufficient light and heat, large evaporation, and scarce precipitation. The observation data of the Ejina meteorological station showed that the annual average temperature was roughly 8.6°C, the difference in daily temperature can be usually 14-17°C and reach the maximum of 34.0°C. The annual average precipitation was 35.00mm and its 60-70% was concentrated during July-September, while the annual average evaporation was more than 3000mm.

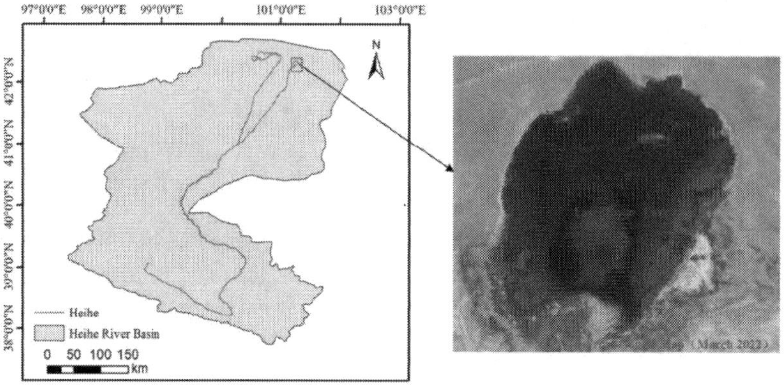

Figure 1. The geographical location of the study area

2.2. Remote sensing data processing

The Landsat satellite data were used. The main tasks of the Landsat satellites were to investigate the earth's resources (such as underground mineral resources, marine and groundwater resources), monitor and assist for rational uses of resources (e.g., agriculture, forestry, animal husbandry, and water conservancy), predict the harvest of crops, explore the general situation of natural geography (e.g., vegetation growth, land use, landform), and forecast various natural disasters (earthquakes, landslides, etc.) or environmental pollution [11]. The basic information of the used satellite data were shown in Table 1.

Table 1. Landsat satellite project properties

System	Launch time	Sensor	Coverage period	Resolution	Number of bands
1	1972	MSS	18 days	80	4
2	1975	MSS	18 days	80	4
3	1978	MSS	18 days	80	4
4	1982	MSS/TM	16 days	80/30	7
5	1984	MSS/TM	16 days	80/30	7
7	1999	ETM+	16 days	Panchromatic 15/multispectral 30	8
8	2013	OLI	16 days	Panchromatic 15/multispectral 30	9

The Landsat satellite data from 2000 to 2017 were employed, with the use of 16-day temporal resolution and 30m spatial resolution. The study area was located within the image range with the row number 134 / 031 or 133 / 031. The freezing period was from December to March of the next year. One scene in January and that taken month by month from April to November were selected, having a total of 144 remote sensing images. The selected images had no cloud coverages in the study area and the ground features were visible.

Landsat data can be downloaded from the NASA website for free (http://glovis.usgs.gov/). The data pre-processing needed radiometric calibration and atmospheric correction of image data. Radiometric calibration was to convert the digital quantization value of an image into physical quantities, such as radiance, reflectivity, or surface temperature and was completed through the radiometric calibration function of ENVI remote sensing processing software.

The acquisition method of remote sensing image data was easily affected by atmospheric aerosol, terrain, and surrounding ground objects, resulting in the interference of the original image containing atmospheric, solar, and other information. Therefore, the atmospheric correction using FLAASH atmospheric correction model was carried out based on radiometric calibration for processing.

IOP Conf. Series: Earth and Environmental Science 1044 (2022) 012009 doi:10.1088/1755-1315/1044/1/012009

3. Result

3.1. Temporal variation of lake surface area

It can be seen from Figure 2 that since the ecological water transfer project was launched in 2000, the area of East Juyan Lake had shown an upward trend. The largest area appeared in 2017, reaching 70.1km^2; the smallest area appeared in 2004, only 10.8km^2. The difference between them was 59.3km^2. From 2002 to 2006, the rising trend of the lake area was obvious, and from 2007 to 2013, this rising trend tended to be relatively gentle. The upward trend was obvious from 2014 to 2017. Before 2003, the area of East Juyan Lake was largely influenced by the water amount entering the lake, and showed obvious periodic change characteristics. After 2004, the seasonal change of lake areas became regular. The water surface area remained stable in winter (December-February in next year), and began to decrease after April reaching a minimum in June or July in a year; after September, the lake area gradually increased and reached a maximum value from October in a year.

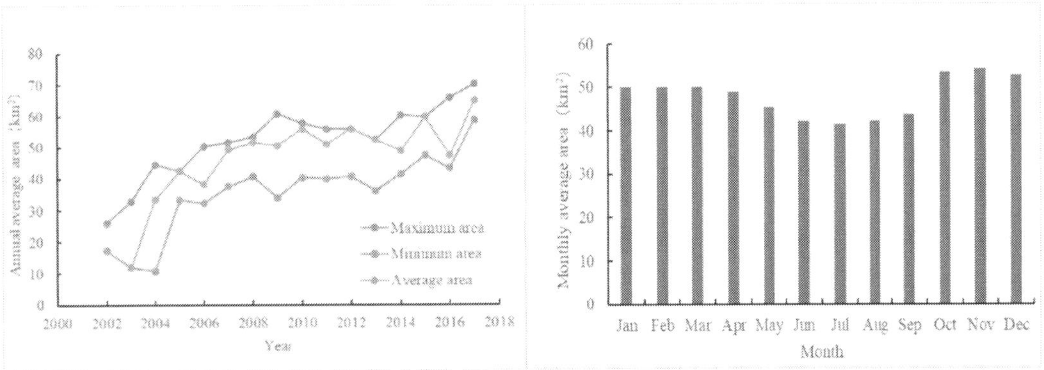

Figure 2. Interannual and intra annual changes in the East Juyan Lake areas

3.2. Temporal variation of lake areas

Figure 3 displays the interannual changes in the spatial distributions of East Juyan Lake. From 2002 to 2009, the lake surface boundary continued to expand. Before 2005, it presented the main expansion of the southern boundary. After 2006, the expansion of the East-West lake surface boundary was more obvious. After 2010, the boundary generally remained stable, with only small changes in the south and east-west directions. On the whole, the water surface area of East Juyan Lake had been increasing to a stable situation because of the water conveyance project implemented in the Heihe River.

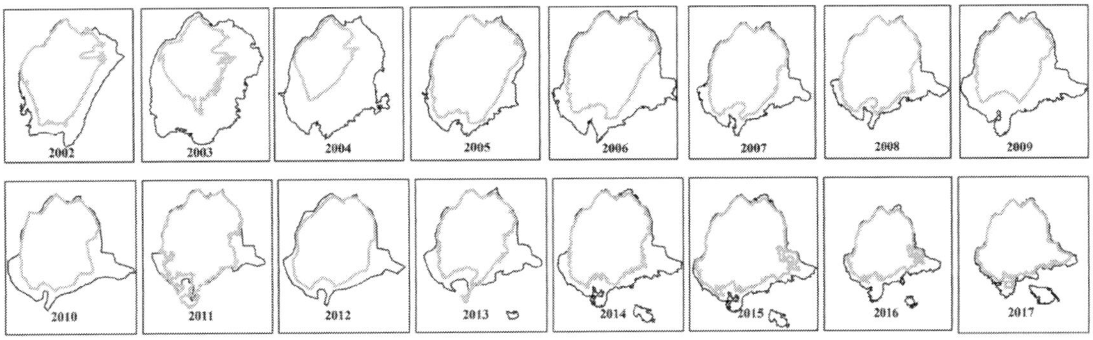

Figure 3. Comparison of minimum and maximum water surface area from 2002 to 2017

4. Conclusion

In this paper, the water surface areas of East Juyan Lake were analyzed using Landsat remote sensing data from 2002 to 2017. The main conclusions are as follows:

There were different interannual variation characteristics of the East Juyan Lake area. In general, the water division to downstream Heihe under the unified water resources management led to a good restoration of lake water bodies and increasing surface boundaries for terminal lakes. The increasing trends were obvious from the annual average of 17.34km^2 to 42.49km^2 during the periods of 2002-2006 and from the annual average of 49.36 km^2 to 55.96 km^2 during the periods of 2014 to 2017. These upward trends became slight from 2007 to 2013 from the annual average of 47.34km^2 to 64.81km^2.

For the overall seasonal characteristics, the water surface areas of East Juyan Lake in winter were almost the largest in a whole year, with an annual average of 51km^2, followed by that in autumn (50.45km^2). The annual average value of water surface area in spring (48.16km^2) was larger than that in summer (41km^2). The lake areas began to shrink from April, reached the minimum in June or July, and increased after September (reaching the maximum from October).

For spatial changes, the lake surface area generally increased from 2002 to 2009, mainly expanded in the south direction before 2005, changed significantly in the eastern and western boundaries after 2006, and tended to be gentle after 2010 with only small changes in the south and east-west directions.

Studies on how to improve the water use efficiencies, suitable wetland scale, and landscape values for lakes are of great importance, which would need to be further explored in the Heihe River for the future.

Acknowledgments

This study was funded by the Basic Scientific Research Special Fund of Central Nonprofit Research Institutes (HKY-JBYW-2020-04).

References

[1] Liu Y, Yang J, Chen Y, Fang G, Li W. The temporal and spatial variations in lake surface areas in Xinjiang, China. Water, 2018, 10:431.

[2] Zhang M, W S, Gao G, Fu B, Ye Z, Shen Q. Exploring responses of lake area to river regulation and implications for lake restoration in arid regions. Ecological Engineering, 2019, 128: 18-26.

[3] Candela L, Elorza FJ, Tamoh K, Jiménez-Martínez J, Aureli A. Groundwater modelling with limited data sets: the Chari-Logone area (Lake Chad Basin, Chad). Hydrological Processes, 2014, 28: 3714-3727.

[4] Khan S, Khan I, Zhao M, Chien H, Lu Q, Ali MAS, Khan AA, Fahad S. Spatial heterogeneity of ecosystem services: a distance decay approach to quantify willingness to pay for improvements in Heihe River Basin ecosystems. Environmental Science and Pollution Research, 2019, 26: 25247-25261.

[5] Liu X, Yu J, Wang P, Zhang Y, Du C.Lake evaporation in a hyper-arid environment, Northwest of China-measurement and estimation. Water, 2016, 8:527.

[6] Xiao S, Peng X, Tian Q. Climatic and human drivers of recent lake-level change in East Juyan Lake, China. Regional Environmental Change, 2016, 16: 1063-1073.

[7] Yang L, Yue L, Li Z. The influence of dry lakebeds, degraded sandy grasslands and abandoned farmland in the arid inlands of northern China on the grain size distribution of East Asian aeolian dust. Environmental Geology, 2008, 53: 1767-1775.

[8] Yu T, Si J, Feng Q, Xi H, Chu Y. Simulation of pan evaporation and application to estimate the evaporation of Juyan Lake, Northwest China under a hyper-arid climate. Water, 2017, 9:952.

[9] Li B, Zhang Y, Wang P, Du C, Yu J. Estimating dynamics of terminal lakes in the second largest endorheic river basin of Northwestern China from 2000 to 2017 with landsat imagery. Remote Sensing, 2019, 11:15.

[10] Shen Q, Gao G, Lü Y, Wang S, Jiang X, Fu B. River flow is critical for vegetation dynamics: lessons from multi-scale analysis in a hyper-arid endorheic basin. Science of the Total Environment, 2017, 603: 290-298.
[11] Xu F, Li H, Bao H. Performance comparisons of land institution and land regulation systems on water area decrease. Habitat International, 2018, 77: 12-20.
[12] Zhang Z, Zheng Y, Han F, Xiong H, Feng L. Recovery of an endorheic lake after a decade of conservation efforts: mediating the water conflict between agriculture and ecosystems. Agricultural Water Management, 2021, 256:107107.
[13] Cao H, Han L, Liu Z, Li L. Monitoring and driving force analysis of spatial and temporal change of water area of Hongjiannao Lake from 1973 to 2019. Ecological Informatics, 2021, 61:101230.
[14] Zhang Z, Chang J, Xu C, Zhou Y, Wu Y, Chen X, Jiang S, Duan Z. The response of lake area and vegetation cover variations to climate change over the Qinghai-Tibetan Plateau during the past 30 years. Science of the Total Environment, 2018, 635: 443-451.
[15] Emami H, Zarei A. Modelling lake water's surface changes using environmental and remote sensing data: A case study of lake urmia. Remote Sensing Applications-Society and Environment, 2021, 23:100594.
[16] Zhang C, lv A, Zhu W, Yao G, Qi S. Using multisource satellite data to investigate lake area, water level, and water storage changes of terminal lakes in ungauged regions. Remote Sensing, 2021, 13, 3221.
[17] Wang J, Sheng Y, Tong, T. Monitoring decadal lake dynamics across the Yangtze Basin downstream of Three Gorges Dam. Remote Sensing Environment, 2014, 152, 251-269.
[18] Yue H, Y Liu, Wei J. Dynamic change and spatial analysis of Great Lakes in China based on Hydroweb and Landsat Data. Arabian Journal of Geosciences, 2021, 14: 149.
[19] Zhang G, Yao T, Chen W, Zheng G, Shum C, Yang K, Piao S, Sheng Y, Yi S, Li J, O'Reilly C, Qi S, Shen S, Zhang H, Jia Y. Regional differences of lake evolution across China during 1960s-2015 and its natural and anthropogenic causes. Remote Sensing Environment, 2019, 221: 386-440.
[20] Nitze I, Grosse G, Jones B M, Arp C D, Ulrich M, Fedorov A, Veremeeva A. Landsat-based trend analysis of lake dynamics across northern permafrost regions. Remote Sensing, 2017, 9: 640.
[21] Babaei H, Janalipour M, Tehrani N. A. A simple, robust, and automatic approach to extract water body from Landsat images (case study: Lake Urmia, Iran). Journal of Water and Climate Change, 2021, 12: 238-249.

IOP Conf. Series: Earth and Environmental Science 1044 (2022) 012010 doi:10.1088/1755-1315/1044/1/012010

Green Architecture as A Facilitating Mechanism for Regional Development and Sustainable Cities-The Case of the Island of Chiloé, Chile

Pareti, S[*1, a], González, V[2, b], Flores, D[3, c] and Pareti, M[4, d]

[1] Andrés Bello University, Economics Department
[2] Federico Santa María Technical University, Architecture Department
[3] Pontifical Catholic University of Chile, Architecture Department
[4] Pontifical Catholic University of Valparaíso, Architecture Department

[a, *] pareti@uandresbello.edu; [b]viviana.gonzalez@sansano.usm.cl; [c]daflores2@uc.cl; [d]martinapareti@gmail.com

Abstract. The purpose of this study is to explore how to develop a sustainable architecture to recover the vernacular architecture, and through this strengthen the socioeconomic development of a city It is proposed that through these recoveries and promotions, the city can welcome sustainable logics, also considering its geographical advantages, natural resources and bordering the sea as characteristics to take advantage of. For this, the island of Chiloé in southern Chile has been selected as a case study, due to (1) it has a unique vernacular architecture, the palafitos (stilt houses); and (2) It has a geographic morphology that has allowed, from its genesis as a city, to build in wood and other sustainable materials, consequently, this can be used as a basis for achieve a restoration of its communities respecting the interrelation of various social, cultural and environmental factors and whose sustainability is established by the balance of the previously mentioned factors.

1. Introduction

The purpose of this study is to explore how to develop a sustainable architecture and recover the vernacular, strengthening the socioeconomic development of a city, making it sustainable considering its geographical advantages, natural resources and coast. For this, the Chiloé Island region of Los Lagos, Chile is taken as a case study. The Island stands out for its ancestral and patrimonial architecture, which remains almost intact, since the arrival of the Jesuits during the seventeenth and eighteenth centuries, in their circular missions, leaving a mark on the culture of the town, as well as the conservation of culture and religious tradition. Currently, the palafitos (vernacular architecture) have undergone structural modifications due to the high tourist demand in the area, eliminating spaces such as the fire pit and adapting ceiling heights. Therefore, they were rescued and explored, for the analysis of the fusion between vernacular architecture and contemporary green architecture (sustainable). Translated into a living and restored heritage in the Chiloe culture. It is concluded that focusing on developing a sustainable and flexible architecture strengthens the socio-economic development of a city, making it sustainable and resilient, as has occurred in the case study of the Island of Chiloé.

2. Theoretical framework

The theoretical framework of this study focuses on two points. On the one hand, the theory regarding vernacular architecture is analyzed, followed by sustainable rehabilitation.

Vernacular architecture is the defining architectural form and style of a region or area, where its architecture is defined by the knowledge and experience of the inhabitant, in addition to the use of local materials, and makes it possible to be in a constant evolution by new forms of construction. The definition of "palafitos" would be houses built on stilts, almost always made of wood, which supports the entire building. Most of these constructions in the world are located in lagoons, rivers, lakes or on the seashore.

In Meilen in 1984, the first stilt house discovery was of great amazement since it was considered that these forms of construction could not be carried out because they were on platforms above the water level [1]

Another clear evidence of these events is that fifty-six of this human colonization were found in Switzerland. These archaeological remains are unique because of their great conservation and interesting culture, they integrate a very important role because they can be used as a study for the agrarian societies of the area [2]

Just as there are also examples of constructions with modifications in their architecture because of the area where they are located, this can be evidenced in the location of Nueva Venecia, Bocas de Aracataca or Buenavista which being close to shore, the foundation of their constructions had to modificate the typology of their palafitos, being a response to having to adapt to the context of the area or region where they were leaving proof of their construction traditions [3].

In case of the vernacular architecture of the palafitos located on the shore of the island of Chiloé, these constructions are posed on stilts, which are built by the inhabitants of the island who, being on the shore of the river or sea, their works are like fishermen or collectors of marine products, since their environment is the condition of shores, but over time these constructions have changed their morphology in order to attract the attention of tourists [4]. Vernacular architecture is developed by its inhabitants depending on the territory in which they are located, which can be linked to the land, sea, nature and its history, therefore, giving a new expression to traditional architecture [5]. Both its heritage and its vernacular architecture is the maximum expression of the identity of a culture of its inhabitants of a region or zone, giving the world its own characteristic [6].

There is evidence of how palafitos have existed since Chiloé originated as a town that began to receive inhabitants, which has made that many of them have adapted to the needs of its inhabitant, as well as technological advances. For this reason, some examples from the most ancient to the more recent of heritage related to palafitos, in areas and regions of different cities. A clear example of vernacular architecture and its identity with the community are the first settlements of buildings in palafitos and were found in sub-Alpinas in countries that today are Germany, Italy, Slovenia, Switzerland and France, which date that they are archaeological finds being one of the oldest from the year five thousand to five hundred before Christ [7].

As the island of Chiloé was removed from the Chilean mainland, its inhabitants have adapted their own culture which is reflected in their own architecture and has taken so much impact that they have adjusted the materiality of the constructions, and giving the vernacular architecture of the Island its external influences, leaving them reflected in its houses built with native wood on the shores, a daily image of the life of the Chilotes [8].

In relation to sustainable architecture, in order to have development, each country has its history where each one has explored different natural resources in order to have a growing economy of their own [9], and the geographical area of Chiloé is no exception. One of the first academic studies is the traditional architecture of Chiloé, which has been extensively analyzed, becoming the most remarkable part of the archipelago.

For sustainable rehabilitation it is necessary to keep its culture alive, so local communities, which are socially and economically linked, must always be active. The history of Chiloé architecture can be maintained in the time necessary to protect its physical constructions, which are characterized by using native wood of the archipelago, but it is essential that they maintain them because of the climatic condition of the area, as well as rain, fungi, humidity and, now in recent times, forest fires [10].

Sustainable architecture is a general term. It is the explanation of design techniques in architecture, which have a vision of ecology for designs, in addition to energy awareness, which makes it adapt to the environment, having modifications regarding its type of construction. Many of these elements can be of vernacular or ancestral architecture, which are adapted to provide a sustainable solution.

According to the article by Masrour and Karbaschi, in their research, they suggest that the real utility of vernacular constructions and technologies should be tested in a critical and systematic way, which may cause modifications or updates for new sustainable buildings in order to reduce the environmental problems of this new millennium, without losing its identity [11].

3. Methodology

The study starts through the analysis of secondary sources, focused on matters of urbanism, vernacular architecture and green architecture, focused on relevant actors within the Chiloé archipelago.

Taking as support the speeches from actors such as the architect and Chilean national prize of Architecture, Edward Rojas, under the ArtificaTuCasa program [5] of the Pontificia Universidad Católica de Chile and organizations such as the Fundación Amigos de las Iglesias de Chiloé, the analysis of the site regarding the Ekistic urban theory proposed by Constantinos Doxiadis [12] regarding human settlements, understanding that the urbanization of the archipelago has always been conditioned by its geography, which implies the understanding of three primary conditions of Chiloé regarding its habitability: the predominance of the sea, the condition of the land as available space, and the shore of the sea as a point of constant interaction between these two worlds [5], therefore we are facing a case of multiple singularities according the urban matter.

The currently analysis follow a scalar line, from the macro scale to the micro scale, therefore, to finalize the research, part of the vernacular architecture of Chiloé will be analyzed, specifically the palafitos, as an example of a green and sustainable architecture that responds to the needs of the site with the available materials, in this case, wood, which also consolidates knowledge and techniques typical of Chiloé carpentry, keys to the future of the place. With this in mind, the analysis will not focus only on the material from the constructive perspective, but also on the immaterial as part of the fundamental traditions of the case study.

4. Results

The urban development of the Chiloé archipelago has always been conditioned by its geographical characteristics, which, although restrictive, gives us a particular type of city that knows how to adapt to its direct context and respond to its needs in a good way. As already mentioned, the habitability of Chiloé is framed in three main territorial conditions: the sea, the shore of the sea and the available land. In addition to this, the cultural syncretism that gave rise to the current Chiloé, plus the insularity that conditioned the construction of human settlements allowed a perfectly recognizable architectural tradition that is repeated throughout southern Chile, where, as has already been said, wood is the undisputed protagonist.

Ekistics, as a disciplinary vision that intends to understand human settlements at their different scales, also uniting the physical with the social dimension [12]-[13], proposes that the success of a human settlement depends on the balance between five ekistics units: nature, man, shelters, society and networks. At an urban level, the archipelago itself is nature, despite the obvious, man is the chilote, who, as an inhabitant, becomes society, connected by two primordial networks. The first one is social, understood as an own identity formed over the years and framed by the conditions of living in an insular environment, and the second, physical, in reference to the multiple roads, not only terrestrial, but also maritime, being understood as the main part of the studied habitat, without going further, are the arteries that allow mobility in this place.

From the beginning of the territorial expansion and the creation of the first settlements, always conditioned by the isolation that insularity implies, it was necessary to create a model of sustainable growth, based on the available resources and the knowledge of the people. The immediate response was wood, given its abundance and physical capacities to solve both the inclemency of the terrain and the climate. The cultural syncretism between the native peoples and the Jesuits also becomes fundamental by complementing the ways of working with the material, creating its own style of carpentry that today goes beyond the way of building, but is transformed into an identity stamp of the place almost as a registered trademark. It is a fundamental part of an architectural typical style of the islanders that also forms an image of a recognizable city, which strengthens social networks and allows the concretion of a way of living and a clear vernacular architecture.

The vernacular Chilote does not only stand out in relation to Green architecture from the material view, understanding the relevance of the low carbon footprint of wood, and its thermal behavior understanding its relevance for the climate of the place, in addition to traditional construction techniques. It also stands out from its social role, understanding that the consolidation of this type of architecture is also connected to a cultural heritage formed over years, where religion and the relationship with the sea are fundamental for its development. Although the recognition of vernacular models is not new from a material perspective, since the efficiency of each one of its geographical area has been proven, and so validating ancestral knowledge, the combination of the inmaterial within its dynamics enriches the revalidation process by understanding that the construction of a specific community and its growth is part of a sustainable development. A clear example of this are the Chilote palafitos (figure1), where the material and immaterial coexist to make a sustainable social and urban development.

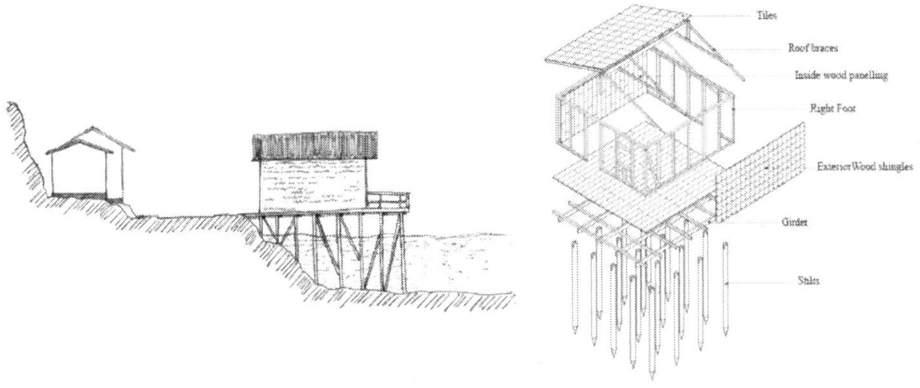

Figure 1. Stilt house sketch. Own elaboration

The dynamics of the shore of the Chilote sea, based on the multiple tidal cycles, added to productive activities such as fishing, promoted the typical construction typology of the island (figure 2), the palafitos, which, together with the churches and the larch tile, are the predominant icons of the archipelago's imaginary. The validity of the firsts ones and their closer relation with the notion of inhabiting, also related to the ekístic analysis model. From the beginning it allows an analysis from the microscale, understanding the architectural element as an example of sustainable architecture.

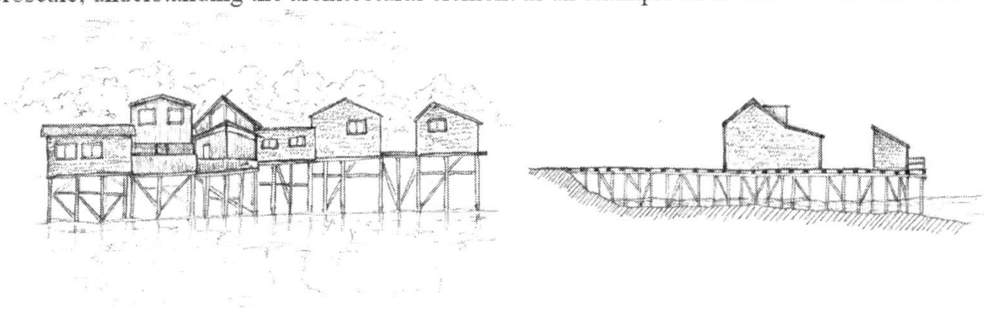

Figure 2. Elevation sketch of stilt houses showing different variations. Own elaboration.

About the conjunction of the material and the immaterial, the economic factor also features, specifically related to tourism, where the iconic appeal of the palafito has implied changes in its uses, so it is not only about houses as in yesteryear, tourist equipment today also takes part. Different schools of thought could consider this phenomenon positive or negative, regardless of this, the palafitos as an element has managed to cope with the reuse thanks to committed professionals who, understanding its heritage value, work according to what that implies [5].

The notion of working with what already exists related to the "sustainable rehabilitation" mentioned within the theoretical framework and tourism as a changing factor of certain dynamics of the archipelago, also implies another focus for a sustainable development of the city and green architecture, within a context where the available land is limited. The reuse of existing buildings thinking of an introspective rather than expansive urban development [14]-[15] allows the conservation of natural green areas that in other contexts would be depredated by real estate developments, for example. This also protects the development of specific flora and fauna that are a fundamental part of the archipelago among other benefits, which will not be furthermore studied in this research. The palafito is presented as a clear example of this urban introspection, being the object of reuse instead of disposable architecture.

In conclusion, the palafito from the material logic is presented as a key object of a vernacular architecture that has always gone hand in hand with the sustainability of its context in accordance with the contextual implications of its time and its technological and social changes. It not only stands out for being a typical wooden construction, with ancestral techniques, but also from the relation with its location, responding to the need for useful land and the productivity of the inhabitant. Socially and culturally, it stands out as a key element of an ancestral tradition, born of cultural syncretism where indigenous and European knowledge falls, which are the basis of the Chiloe idiosyncrasy, where its understanding is key to the urban evolution of the archipelago, understanding that when we talk about sustainability, it is not only about the physical aspect of what has been built. Sustainable development should also be taken care of the cultural heritage of each city.

5. Conclusion

The current study explores the rescue of vernacular architecture and its union with sustainable architecture, in order to have a dialogue between the tradition of the Chiloe culture, in this case presented in the palafitos and the modernity, that would be the modifications inhabitants have suffered. So much, that they have transformed their cities to become landscapes of urban tourism that can be enjoyed, this being sustainable architecture, giving a new architectural language.

In the analysis of the Ekistic urban theory proposed by Constantinos Doxiadis [12] regarding human settlements, it is implied that the urbanization of the Archipelago has always been linked to its geography, considering three primary conditions of Chiloé regarding its habitability: the predominance of the sea, the condition of the land as an available space, and the shore of the sea as a point of constant interaction between these two worlds [5]. In response, new urban construction models are needed, where heritage is preserved and new practices adapt to the Chiloe culture with strength.

The Chiloé archipelago has always been aware of its context and its particularities, mainly geographical, responding to its needs according to its capacities regarding what habitability leads to. The expertise in the use of wood, responds to the impossibility of bringing materials from the continent and its productivity capacity, which, although today is regulated to protect tree species such as larch, they led to the creation of a personal architectural style and sustainable within its conditions, understanding the ephemerality of wood regarding external agents such as humidity and pests. We do not speak only of a material aspect, but also of the consolidation of our own identity reflected in this type of architecture, exemplified in the palafitos as an architectural element.

This same consolidation, both material and immaterial, should be a fundamental part of the conception of green architecture, considering the present and the future, understanding that the construction of sustainable cities should be consolidated with representative communities in pursuit of

IOP Conf. Series: Earth and Environmental Science 1044 (2022) 012010 doi:10.1088/1755-1315/1044/1/012010

particular and culturally rich models, assuming that sustainability cannot be standardized, but must respond to the needs of each specific context.

References

[1] A A Palau 1952 *¿Los palafitos fueron lacustres?* Empúries: revista de món clàssic i antiguitat tardana, pp. 190-193

[2] UNESCO 2011 Palafitos del entorno de los Alpes [Online]. Available: https://whc.unesco.org/es/list/1363#. [Last access:November 14, 2021]

[3] J J Gómez 2014 *Vivienda en hábitats lacustres* [Online]. [Last access: November 14, 2021]

[4] P R Campos 2021 *Arquitectura vernácula en palafitos de Chiloé* Building & Management, vol. 5, nº 2

[5] E. Rojas 2021 *Artifica tu casa 2021: Edward Rojas habla sobre la arquitectura de Chiloé* [Online]. Available: https://www.youtube.com/watch?v=WY2ZLPGSCdg. [Last access: November 14, 2021]

[6] I C I S T CULTURAL, *ICOMOS* 2021 [Online]. Available: https://www.icomos.org/charters/tourism_sp.pdf. [Last access: November 14, 2021]

[7] A Bahamón and A M Álvarez 2009 *Palafito: de arquitectura vernácula a contemporánea*, Barcelona: Parramón

[8] P Martínez 2018 *El bordemar en la Isla de Chiloé* ETSAM, Madrid

[9] J Arwel Edwards 1996 *Mines and quarries: Industrial heritage tourism* Annals of Tourism Research, vol. 23, nº 2, pp. 341-363

[10] M. Hurtado Saldías 2018 *Metodología para una rehabilitación arquitectónica sostenible: El caso de los palafitos de Chiloé* Arquitecturas del Sur, vol. 36, nº 53, pp. 36-57

[11] M J K and M M Masrour 2015 *Comprehending Vernacular Architecture in the Architectural Educational System in Iran* American Journal of Civil Engineering and Architecture, vol. 3, nº 2, pp. 45-51

[12] C Doxiadis 1970 *Ekistics, the Science of Human Settlements* Science, vol. 170, nº 3956, pp. 393-404

[13] L Galindo 2009 *Ekística y transformación de ciudades* Universidad del Rosario, vol. 4, nº 14

[14] O Bohigas 1985 *Reconstrucción de Barcelona* MOPU, vol. 62

[15] F Carrión, 2005 *El centro histórico como proyecto y objeto de deseo* Revista Eure, Santiago Pontificia Universidad Católica de Chile, vol. 31, nº 93, pp. 89-100

A Conceptual Framework for Developing and Transitioning to a Green Economy in Sub-Saharan Africa: A Focus on Nigeria

Alfred John Ogbonna[*1, a, b]

[1]School of Economics and Management Southeast University, Nanjing, China.

Corresponding Author: [*, a]233159951@seu.edu.cn; [b]rabbialfred@yahoo.com

Abstract. Green Economy (GE) is among the sustainability trends meant to tackle challenges in the present norm. As such, for developing countries in Sub-Saharan Africa (SSA) to indulge and transform their economic activities in line with GE standards, there is a need for a workable and pragmatic framework for GE that the stakeholders agree and accept. Hence, this research identified and examined GE-related issues and developed a conceptual framework for a green economy to examine the GE related factors and stages in transitioning to GE in Nigeria. The framework identified and incorporated four stages: the formation, implementation, results and post GE stages by examining the factors at every stage. An exploratory desk study was adopted to analyze the secondary data to examine the linkages in factors that promote/accelerate or hinder/impede GE concepts. The research highlighted the framework's feasible application and adaptability in Nigeria.

1. Introduction

Recognizing the quest for social and economic development accompanied by environmental challenges that aids the deterioration of our planet's ecosystem has resulted in a global consensus to adopt sustainable and environment-friendly approaches. Such concepts, principles, approaches are corrective or preventive measures to enable overall sustainability in all spheres of human activity, as most economic activities are mostly towards achieving better living standards; with moderate or no damage to the environment. To achieve a green economy (GE), a framework acceptable to all stakeholders, from policymakers and industrialists to professionals and the general public, is required. Past and recent studies and also reports[1],[2],[3],[4] emphasized that global consistency is important when it comes to dealing with environmental issues that will eventually lead to harmony. This includes balancing ecosystems, preventing climate change, and ensuring a sustainable and healthy built environment (BE) while ensuring the socioeconomic and socioenvironmental development of humans on Earth. As a whole, these factors led to the UN General Assembly's approval of the 17 Sustainable Development Goals in 2015. (SDGs). That's why the green economy (GE) concept came about. It's a way to make sure businesses, economies, and the built environment are sustainable while not stopping or slowing down the human race's social and economic development (SED)[5].

GE refers to the whole economy being eco-friendly, sustainable, and resilient in all sectors typically linked with environmental products and services, such as renewable energy and other clean technology[5]. GE concepts advocates measures to mitigate ecological, environmental, and resource risks and scarcities only to achieve sustainable development (SD) without causing harm to the ecosystem or the environment[6],[7]. It is a concept that improves the quality of life for humankind while maintaining the planet's ecological boundaries [8]; the concept optimizes value and growth throughout the economy while advocating for optimum management of natural resources sustainably[9]. It is wise to say that the GE concept is socially inclusive, resource-efficient, and environmentally friendly. Public and private investment in key economic activities, green portfolio of assets, and infrastructural projects that reduces carbon emissions and pollution and improves energy and resource efficiency helps keep biodiversity and ecosystem services for humans from being lost. [10].

According to the 2011 UNEP GE Report, "for an economy to termed green economy, it must be fair and efficient; fairness means that nations must address global equity concerns by ensuring a just and equitable transition to a low-carbon, all resource-efficient, and socially inclusive economy" [11]. Inclusive GE contributes to human well-being and social fairness while mitigating environmental hazards and resource scarcity. GE is a viable alternative to the current dominant economic paradigm, which exacerbates disparities, promotes resource waste that results in scarcity, creates broad hazards to human health, and damages the environment[12]. It means that GE is connected to ecological economics but does have a more policy-oriented orientation[6],[7]. As a result, UNEP began promoting GE ideas to policymakers and regulators whose influence spread across all sectors and industries.

For developing countries in Sub-Saharan Africa (SSA) to indulge and transform their economic activities in line with GE standards, there is a need for a workable and pragmatic conceptual framework for GE that is agreed upon and accepted by the stakeholders. A framework is a support structure or system that connects ideas, elements, steps, and processes as the primary structure. A conceptual framework is a textual or visual model depiction and predictions of all variables' relationships. In general, the conceptual framework is designed based on a review of previous research and ideas on the issue. A workable and pragmatic conceptual framework for GE in SSA countries must or should entail balancing the needs, interests, and requirements of the government ministerial and environmental agencies, industrialists, environmental professionals, academia, and the environmentalist pressure and other concerned individuals/groups.

However, Some recent studies[5], [13]–[17] have argued and emphasized that in light of the current economic crisis and the perception that sustainability policies and measures cannot be easily implemented successfully, politicians have placed their hopes in greening the economy. However, there are substantial flaws with this concept's aims and approaches. Green economy policies, in particular, will fail to fulfil their claimed goals and objectives of reversing environmental degradation and reducing poverty if political, economic, and cultural constraints are not fully considered and addressed.

Hence, this research identified and examined GE related issues and developed a conceptual framework for a green economy in SSA countries like Nigeria with the views and aims of examining the GE success factors, the GE hindrance/barrier factors, the GE promoters/drivers based on the interests and impacts on the concerned stakeholders involved in transitioning to GE in Nigeria.

2. Methodology

The research adopted exploratory concepts based on a desk study methodology that will analyze only secondary data. Desk study is conducted only through research rather than physical investigations while sitting at a desk. It also provides early knowledge of a topic or situation, highlights possible dangers, and drives further investigations' scope, specificity, and approach. Desk research is mostly concerned with gathering secondary data from existing resources; hence, it is a low-cost strategy compared to field research since the direct costs are internet costs, executive time, phone rates, and directories. Desk research is quite successful and may be undertaken in the early stages of a trend or transition research (especially the current GE trans since it is fast and inexpensive, and most essential information can be quickly obtained and utilized as a benchmark in the research process[18], [19]. Secondary data analyses entail reviewing previous research studies and reports, often disclosing much historical data or decision-making tendencies. These will aid in identifying major successful and failed approaches. Solutions that did not gain attention in the past due to various factors are exposed as viable issues for additional research[18]. Figure 1 below shows the methodological steps of the study.

Figure 1. Research Methodological Framework.
Source: Authors' Illustrations

3. Literature Review

Like the climate change and the climate economics debate, the GE issue seems considerably different from the inside than it does in the media. Any novice or casual reader may conclude that there are valid worries about lowering emissions without negatively impacting the economy. After reducing the noise created by special interest groups, environmental economists think that a market-based programme to tackle climate change might achieve significant improvements at a low, if not insignificant, cost. There is less agreement on how rapidly we should go, whether massive conservation programmes should begin immediately or be phased in over time[20]. Climate policy economists agree on certain fundamental points. Many people agree that we need to place a price on all carbon emissions and that the adverse economic repercussions of this policy will be bearable. That is, we should indeed act to slow global warming. Experts disagree on the timing of a big hike in carbon costs.

Several studies [2] [5]–[9], [13]–[17] have identified, examined, and discussed issues related to GE, green growth, green production and consumption in this century. Their findings show that the green economic ideology is based on the scientific concept of eco-centrism, which recognizes the biosphere as the origin of life and the need for the environment and civilization to coevolve. The essence of this category depicts the union of two components: the environment and the economy, which form specific integrity within the framework of a holistic approach. However, this requires transitioning to GE in most spheres of the economy across various sectors with buy-in from the major stakeholders.

3.1 Transitioning to a Green Economy

In order to comprehend the uneasy balance of transitioning from conventional to GE and Sustainable Development, Bina[21] examined three interconnected patterns: scarcity and limitations, means and ends, and reductionism and unity, which increase the knowledge of the conflicts between new proposals. The patterns help explore and clarify the implications of GE for sustainable development by revealing an economization and polarisation of discourses, a persistently flawed interpretation of sustainable development, and a conflict between the fixing or changing of dominant socio-economic paradigms perspectives that underpin its conceptualization.

According to Ogbonna J.A.[5], in some developed countries, a feasible GE may be less difficult, more demanding, and present fewer hurdles, but in SSA developing countries such as Nigeria, the opposite is true. From 2016 to 2018, the Global Green Economy Index (GGEI) placed Nigeria 53rd out of 130 nations, while the 2019/2020 reports put Nigeria 56th internationally and 19th in Sub-Saharan Africa, despite being Africa's biggest economy. This led to Nigeria receiving a very low GGEI grade, suggesting that the country's GE transitioning initiatives are more or less slow or may have remained stagnant. These were due to various factors that might either accelerate or impede the transitional journey towards GE.

GE encompasses several components, steps and processes. Barbier[22], Krugman[20], and Altenburg et al.[13] stressed that Eco-friendly growth, energy efficiency, and renewable energy are all complimentary in the eyes of international policymakers. Pollution and global ecosystems result in the loss of various benefits, or "services," since these systems are used for human usage and activity. Concerns regarding the environment and the economy must be addressed. The goal is to overcome commercial, political, and institutional shortcomings that obscure the economic value of scarcity. We must first pay for the maintenance and conservation of ecosystems to preserve and secure them. Understanding ecological scarcity necessitates a more thorough economic and scientific examination of lost benefits, as well as policy ramifications. Taxes on financial and currency activities, international lending facilities, and other creative funding options may be beneficial [22].

3.2 Stakeholders Roles in GE Transition and Development

The emergence of GE signalled a transition to sustainably greening production and consumption. As such, the effectiveness of this transition requires the buy-in and cooperation of the major stakeholders. These include the government ministerial/environmental regulatory agencies that play a role in policies; the industrialists who are at the center of manufacturing; the professionals who signal trends; the academia who research on trends; the environmentalist pressure groups whom are the green populist and sound the alarm and other concerned individuals/groups who raises concern when necessary.

Some studies[13]–[16], [20]–[23] indicated that building a GE requires policy changes from the political leadership by countering major challenges for GE and sustainable economic development to ensure effective green industrial policies. Such policies are directed towards accelerating structural change towards Green and inclusive Prosperity that deliver on poverty reduction, the well-being of humankind, low carbon emission, balanced ecosystem while maintaining a great deal of technological innovation during the transition. These studies asserted the notion that building green economies leads to wealthy green economies.

However, there is a paradox here, SSA countries like Nigeria are still trying to industrialize to cater for their needs during such a transition. Some GE processes may affect some industries, generating opposition from the affected quarters. A workable and pragmatic conceptual framework for GE in Nigeria and other SSA countries should embrace the needs, interests, and requirements of key players in major economic sectors by structurally balancing the impacts of the GE concepts on such.

3.3 GE Related Factors: Success Factors, Accelerators / Promoters, Barriers / Hindrances

Policies, concepts, strategies and any other planned or unplanned action usually affect or impact the environment in which they exist. Such effects/impacts range from political, economic, social, technological, environmental, legal aspects or issues within the said environment. In some cases, synergistic effects such as socio-economic and the techno-economic impacts of the GE concept will bolster its perception within the economic environment.

GE has become a priority for many countries and various levels of governments from the last decade onwards. These countries are ready to tackle the century's biggest challenges, from urbanization and resource shortages to climate change and economic instability. To achieve these goals, GE policies are required. Governments, business executives, experts, academia, and NGOs must all be on board to achieve GE goals. All stakeholders must be included in effective GE policies and goals. Organizations conduct independent evaluations and assessments like the Global Green Economy Index (GGEI). Furthermore, analyzing a country's GE performance utilizes all green statistics. The GE grading is heavily influenced by its sustainable consumption and waste production [24]–[29].

As the focal point of the study, Nigeria has a population of 200 million, and a GDP of $500 billion needs to diversify its economy from oil dependent to green, resilient and inclusive economy to achieve the 17 SDGs. These will involve green infrastructure (roads, rail, power, ICT, etc.), urban renewal and greening of its major cities. Sustainable funding is essential to support infrastructural development projects (IDPs) long-term success. IDPs need policies that address the built environment's social, environmental, and economic aspects[4], [5]. Nigeria has also issued a green bond (GB) to help climate change efforts. Nigeria's GB must meet current green bond standards. Nigeria's GE will set a tremendous precedent for other SSA countries. Leadership and political determination are essential to implement GE policy. Synergies in GE research and development will be enhanced via academic and industry collaboration[5].

GE needs adequate resources to promote green projects. Sustainable Infrastructure Finance (SF) helps build infrastructure. SF examines socially responsible investments by including environmental and social factors into financial choices. However, developing countries like Nigeria need international money, collaboration, and FDI to complete GE projects. Globalization facilitated cross-border investments. Sustainable funding, FDI, foreign support, and partnership for GE. Local capacity and assimilation; Socio-economic Pressures and Development; Synergy of GE R&D. These were determined by the study's eleven GE determinant factors[20], [22].

4. The GE Conceptual Framework

The proposed GE conceptual framework encompasses the success factors, the GE hindrance/barrier factors, the GE promoters/drivers based on the interests and impacts on the stakeholders involved in transitioning to GE in Nigeria. The framework consists of four stages: the formation and transition stage (FTS), the adoptability and implementation stage (AIS), The end-result stage (ERS), The post-GE stage (PGES). Figure 2 below outline the framework.

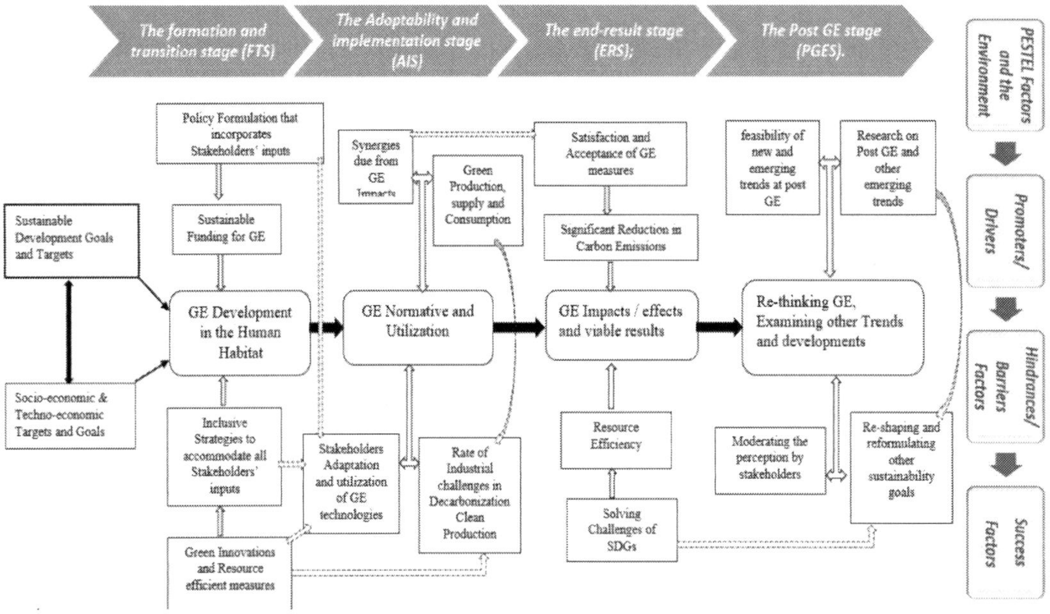

Figure 2. A Conceptual Framework for Transitioning and Developing a GE for SSA countries Like Nigeria

Source: Authors' Illustrations

The framework in figure 2 above consists of processes divided into four stages: the formation and transition stage (FTS); the adoptability and implementation stage (AIS); The end-result stage (ERS); The post-GE stage (PGES). It also encompasses the success factors, GE hindrance/barrier factors, and GE promoters/drivers based on the interests and impacts on the stakeholders involved in transitioning to GE in Nigeria.

The framework shows that the adoption and timely application of a new economic model and a different approach to green development would need the restructuring of SSA nations, such as Nigeria, with sustainable funding and investments and a change in the mode of operation. To guarantee green manufacturing, sustainable supply chain management, and responsible consumption, the framework and the GE developmental strategy must encompass the strengths and limitations of regions, cities, and municipalities. Other factors should take efficient service delivery into account; strong synergy between political, social, economic, and technological advancements that promote economic development and generate employment must be an important goal of the GE framework.

Such an economic shift is a significant development that will need the participation of all stakeholders: particularly the government, industrialists, financial institutions; Academia, professionals, and citizens in all spheres of life; in formulating an effective and inclusive GE policy. Some components of the GE policies will be completely accepted and promoted by stakeholders, while others will be

IOP Conf. Series: Earth and Environmental Science 1044 (2022) 012011 doi:10.1088/1755-1315/1044/1/012011

hindered and hampered due to the good or negative repercussions of such changes on them. The success or failure of GE is determined by how elaborately it is designed and how well its pragmatic implementation tactics and strategies are implemented. These demonstrate the need of tight collaboration with business and civic society since the survival of economic growth is dependent on active community participation and the development of a strong public-private partnership. Table 1 below shows the significant relationship between the processes within the stages and related factors/impacts.

Table 1. Cross Tabulation matrix showing the relationship of key GE stages and Factors

Green Economy Stages	**Green Economy Related Factors/Effects/Impacts**			
	Success Factors	*Hindrances/Barriers Factors*	*Promoters/Drivers*	*PESTEL Factors and the Environment*
The Formation and Transition stage (FTS)	Steering innovations towards a green economy and well-being of mankind	Active or Passive strategies that balanced the needs, interests and requirements of stakeholders and GE issues	Sustainable finance and funding to transition to GE	Policy Formulation with inputs from all key stakeholders; Leadership and political will to ensure GE policies across
The Adoptability and Implementation stage (AIS)	Adopt and utilize Green Technologies, green measures for GE implementation	Decarbonization challenges from various economic sectors with pushbacks from Industry adoption for Cleaner Production	Green production and consumption	Socio-economic and techno-economic synergies due from GE
The End-result stage (ERS);	Resource efficiency and eco-friendly	Challenges in solving and achieving some SDGs due to other aspects of GE	Zero-carbon emission and or Significant reduction of Environmental waste	Socio-political acceptance and satisfaction of GE as a means to overall sustainability
The Post GE stage (PGES).	Re-shaping and reformulating other sustainability goals	Below average performance of GE due to Negative perception by stakeholders	Research on Post GE trends and or advancing new sustainability trends	Socio-economic and techno-economic feasibility of new trends at post GE stage and their respective

Source: Author's illustrations

5. Conclusions

Good frameworks are adap F Carrión, 2005 *El centro histórico como proyecto y objeto de deseo* Revista Eure, Santiago Pontificia Universidad Católica de Chile, vol. 31, nº 93, pp. 89-100 enough to the business's culture and surroundings. As such, the proposed conceptual framework for transitioning and developing a GE in this research must be workable and implementable for effectiveness. If further options are necessary, they may be provided. This enables the addition of choices without increasing complexity while offering a comprehensive perspective.

A green economy is defined by significantly increased investment in economic sectors that contribute to and strengthen the earth's capacity, reduce ecological and environmental hazards. As the foundation for all public services and economic activity, GE necessitates a developmental state approach to address the growing need to alleviate poverty through targeted wealth transfers, new job opportunities, and improvements in the flow of ecosystem-rated goods and services as access to them. SSA countries like Nigeria require investments. The public service infrastructure will need to be reorganized to allow green projects to contribute more to economic growth and green jobs, increased resource efficiency with less waste and pollution, and considerable reductions in greenhouse gas emissions.

Nigeria's transition to a GE will test the current economic system (which is presently based on oil and gas) to suit rising material consumer requirements. As a result, comprehensive policy and planning will be necessary to ensure resource-efficient, low-carbon, and equitable economic development. While changing consumer behaviour and enhancing end-user efficiency will assist reduce society's environmental effect, innovation and green technology will be important in decoupling material growth from natural resource depletion and promoting green job development. This would necessitate new competencies and research in innovation and green technologies, and a new business model focused on broad participation and sound governance to build investor trust.

To achieve the goals and objectives of the GE, the outlined framework, other models and portfolio of methods and actions are required. The Strategic Framework serves as a road map for Nigeria's low-carbon, equitable economic growth, speeding the country's transition to a GE and paving the way for long-term sustainability.

References

[1] UNWCED, "Our Common Future: A Report of the UN World Commission on Environment and Development. The Brundtland Commission Report," 1987. Available: http://www.un-documents.net/wced-ocf.htm.

[2] M. F. Cordova and A. Celone, "SDGs and Innovation in the Business Context Literature Review," Sustainability, vol. 11, no. 24, p. 7043, 2019.

[3] United Nations, "The sustainable development goals report," 2016. doi: 10.29171/azu_acku_pamphlet_k3240_s878_2016.

[4] S. G. Dalibi et al., "Socio-Economic Performances of Mega Construction Projects (MCPs) in the Light of Sustainable Development of Nigeria's Built Environment," IOP Conference Series: Earth and Environmental Science, vol. 495, no. 1, 2020, doi: 10.1088/1755-1315/495/1/012046.

[5] A. J. Ogbonna, "An Assessment of Determinant Factors for a Sustainable Green Economy in Nigeria," in IOP Conference Series: Earth and Environmental Science, 2021, vol. 793, no. 1, p. 12031.

[6] L. Kahle and E. Gurel-Atay, Communicating sustainability for the green economy. 2013.

[7] UNEP - UN Environment Programme, "What is an 'Inclusive Green Economy'?" https://www.unep.org/explore-topics/green-economy/why-does-green-economy-matter/what-inclusive-green-economy (accessed Feb. 05, 2021).

[8] Green Economy Coalition, "Green Economy: A Submission to the UN Conference on Sustainable Development Zero draft text.," 2011.

[9] UK Government, "Enabling the transition to a green economy.," 2011.

[10] UNEP - UN Environment Programme, "Green Economy." https://www.unenvironment.org/regions/asia-and-pacific/regional-initiatives/supporting-resource-efficiency/green-economy (accessed Feb. 05, 2021).

[11] UNEP - UN Environment Programme, "Towards a Green Economy: Pathways to Sustainable Development and Poverty Eradication - A Synthesis for Policy Makers," 2011.

[12] UNEP - UN Environment Programme, "What is an 'Inclusive Green Economy'?" https://www.unenvironment.org/explore-topics/green-economy/why-does-green-economy-matter/what-inclusive-green-economy (accessed Feb. 05, 2021).

[13] T. Altenburg and D. Rodrik, "Green industrial Policy: Accelerating structural change towards wealthy green economies," Green Industrial Policy, 2017.

[14] M. Dittrich, S. Giljum, S. Lutter, and C. Polzin, "Green Economies Around the World," Implications of resource use for development and the environment, 2012.

[15] Z. C. Siamanta, "Building a green economy of low carbon: the Greek post-crisis experience of photovoltaics and financial' green grabbing'," Journal of Political Ecology, vol. 24, no. 1, pp. 258–276, 2017.

[16] C. Bann, P. B. Anand, U. Iftikar, T. Scott, G. Bouma, and S. Dutta, "Towards Green and Inclusive Prosperity: Building green economies that deliver on Poverty reduction," 2015.

[17] U. Brand, "Green Economy; the Next Oxymoron? No Lessons Learned from Failures of Implementing Sustainable Development," GAIA- Ecological Perspectives for Science and Society, vol. 21, no. 1, pp. 28–32, 2012, doi: 10.14512/GAIA.21.1.9.

[18] D. R. Cooper and P. S. Schindler, Business Research Methods., 12th ed. The McGraw-Hill Companies, Inc., 2014.

[19] M. Saunders, P. Lewis, and A. Thornhill, Research methods for business students., 7th ed. Pearson Education Limited, 2016.
[20] P. Krugman, "Building a green economy," New York Times, vol. 5, 2010.
[21] O. Bina, "The green economy and sustainable development: an uneasy balance?," Environment and Planning C: Government and Policy, vol. 31, no. 6, pp. 1023–1047, 2013.
[22] E. Barbier, "The Policy challenges for green economy and sustainable economic development," in Natural resources forum, 2011, vol. 35, no. 3, pp. 233–245.
[23] C. J. Cavanagh and T. A. Benjaminsen, "Political ecology, variegated green economies, and the foreclosure of alternative sustainabilities," Journal of Political Ecology, vol. 24, no. 1, pp. 200–216, 2017.
[24] S. Fineman and K Clarke, "Green stakeholders: Industry interpretations and response," Journal of Management studies, 1996, Accessed: Feb. 08, 2021. [Online]. Available: https://onlinelibrary.wiley.com/doi/abs/10.1111/j.1467-6486.1996.tb00169.x.
[25] M. A. Cohen and M. P. Vandenbergh, "The Potential role of carbon labeling in a green economy," Elsevier, 2012, doi: 10.1016/j.eneco.2012.08.032.
[26] E. G. Hertwich, "Life Cycle Approaches To Sustainable COnsumption: A critical review," Environmental Science and Technology, vol. 39, no. 13. pp. 4673–4684, Jul. 01, 2005, doi: 10.1021/es0497375.
[27] G. R. Spaargaren, "Sustainable Consumption: A Theoretical and Environmental POlicy Perspective," Society &Natural Resources, vol. 16, no. 8, pp. 687–701, Sep. 2003, doi: 10.1080/08941920309192.
[28] T Jackson, Sustainable consumption – Handbook of sustainable development. 2014.
[29] P. Dolan, "The Sustainability of 'Sustainable Consumption,'" Journal of Macromarketing, vol. 22, no. 2, pp. 170–181, 2002, doi: 10.1177/0276146702238220.

IOP Conf. Series: Earth and Environmental Science 1044 (2022) 012012 doi:10.1088/1755-1315/1044/1/012012

Examining the Success Factors for Sustainable Development in Nigeria's Economic Growth and Development

Alfred John Ogbonna[*1, a,b]

[1] School of Economics and Management Southeast University, Nanjing, China

Corresponding Author: [*,a]233159951@seu.edu.cn; [b]rabbialfred@yahoo.com

Abstract. Sustainable Development (SD) is a concept of socially inclusive approach, resource-efficient, with low-carbon measures bolstered by public and private investment partnerships in economic activities and infrastructural projects that minimizes carbon emissions pollution reductions, improves energy usage, and ultimately prevents the loss of biodiversity within the ecosystem services for humankind. Implementing SD policies is paradoxically challenging for a Nigeria and other African countries due to significant factors affecting the success of such policies and efforts within its economy. This study examined 11success factors (SFs) that affects SD in Nigeria's economy. The results show the huge impact of stakeholders on the success of the SD efforts in Nigeria. In addition, SD policy implementation, enforcement, and adaptability may take years or decades, but once embedded in overall economic sectors, organizational operations, and corporate social responsibilities, as well as the majority of the population's respective activities, the positive social, economic, and environmental sustainability impacts, will be profound.

1. Introduction

The United Nations' overarching paradigm for inclusive growth is sustainable development. From the last quarter of the twentieth century to the present decade, there has been a global recognition of the effects of environmental degradation and ecosystem challenges. Globally, people agree that the quest of social and economic progress comes with environmental challenges and the degeneration of the planet's ecosystem. This means that eco and environment-friendly methods should be used. The application of such concepts, principles, and approaches is intended to provide corrective or preventive measures in all spheres of human activity, given that the majority of economic activities are directed at improving living standards while causing minimal or no damage to the environment. This measure and concept is referred to as "Sustainable development (SD)". The first definition of SD is from the Brundtland's Commission report of the UNWEP in 1987[1], which defines SD as the development that meets the needs of the present generation (in their quest for development) without risking the capacity of future generations to optimally meet their own needs(social, economic and environmental developmental needs)". However, since then, several reports and studies have given SD's new dimensions, concepts, and understanding.

SD in "economic sense" refers to the development of all sectors typically associated with social, environmental and economic products and services, such as consumption, renewable energy and other technology that is good for the planet [2]–[5]. SD advocates measures to mitigate ecological, environmental, and resource risks and scarcity solely to achieve sustainability without harming the ecosystem or the environment. It aims to improve humankind's quality of life while preserving the planet's ecological boundaries[5], [6]; SD maximizes value and growth throughout the social-environmental-economic development while advocating for sustainable management of natural resources[7]. SD is a socially inclusive/integrated, resource-efficient, and relatively low carbon concept that is bolstered by public and private spending in infrastructure and economic sectors that significantly reduces carbon emissions, improves renewable energy and resource efficiency, and eventually discourages the biodiversity loss and ecosystem services for humanity[8].

SD is gaining traction globally from both the developed and the developing nations, and Nigeria as a developing country is not an exception. In order to achieve SD for inclusive economic growth in a developing country like Nigeria, many factors have to be considered which may promote or hinder economic developments. A factor can affect the outcome/result of an event, an intention, a decision, and or a circumstance. A success factor (SF) is any knowledge, skill, trait, determination, attitude, value, or other personal traits required to perform the job or role and differentiate normal performance from superior performance. SFs (in any endeavour, be it economic development, academic, sports etc.) encompasses all of the various internal and external influences that can have an impact on the endeavour/activity potential success[9]–[14].

To successfully achieve SD is difficult and paradoxical in many ways. Because human existence and well-being are inextricably related to the natural environment. Finding a balance between human social, economic and technological needs with ecosystem preservation, on the other hand, is a challenge. Adejumo[9] stressed that previous studies assumed that the Earth's capacity to maintain ecological balance was limitless. This was dismissed by UN reports and several studies[1][15]–[17]; the planet may have an upper natural system limit known as assimilation capacity, "throughput constraint," or ideal level[9]. Hence, this research identified and evaluated relevant SFs necessary for SD requisite for Nigeria's economic growth and development; also examined and categorized such factors as barriers, hindrance or accelerators, promoters, and drivers based on the perception of these factors the concerned stakeholders.

2. Research Scope and Methodology

The research scope is limited to the perception of stakeholders concerned with Nigeria's economic, social, environmental and sustainable development from various government ministries, agencies and organizations. Other organizations such as the UN office in Nigeria, Academia, development consultancy firms, non-governmental/profit organizations, and experienced free-lance professionals were administered the research questionnaire to obtain and evaluate their perceptions.

The Mono Quantitative Method was adopted in this study, leading to a research design using a questionnaire survey for data collection and analyses[18], [19]. Such data was from the literature identification of eleven SFs for SD in Nigeria's economic growth. These SFs were subjected to a pilot study (submitted to 30 experienced professionals in Nigeria) to determine their suitability and applicability in Nigeria's SD efforts. Respondents in the pilot study used a 5-point scale to figure out which SFs could be used in the general survey: strongly agree, agree, neutral, disagree, strongly disagree, and not at all agree.

A wider survey proceeded the pilot survey using the revised questionnaire with a different 5-point Likert scale of impacts: extreme, strong, moderate, low, no impact at all; distributed based on a simple random technique; and the sample size was determined using a Krejcie and Morgan's [20] table, which fixes 384 sample size for upto one million research population size. Quantitative analytical approaches such as Cronbach Alpha for data reliability and consistency, Mean Item/average Score (MIS), Relative/ranking of Importance Index (RII) evaluation models, and T-test statistics were utilized to evaluate the SFs for SD in Nigeria's economic development.

3. Success Factors (SFs) for Sustainable Development (SD) in Nigeria

Globalization has increased the degree of competitiveness and pollution in the environment. The industrialization age has reduced the cost of living on natural resources. The main motivators of sustainability include increasing levels of pollution, rising global temperatures, and decreasing green

IOP Conf. Series: Earth and Environmental Science 1044 (2022) 012012 doi:10.1088/1755-1315/1044/1/012012

areas. Previously, sustainability was seen as an optional component, but it is now deemed crucial for the globe's existence. SD Sustainable production and consumption include using reusable, recyclable materials with minimal energy usage in production to generate a high-quality product that reaches the end-user while causing the least amount of environmental damage. Saving resources is equivalent to generating them. A sustainable manufacturing system creates a balance between the economy and the environment. Environmental responsibilities are mandated by law and give a competitive advantage[10].

The four dimensions of SD are society, the environment, culture, and economics. These dimensions are inextricably tied and therefore cannot be separated. Sustainability is a way of looking forward that considers environmental, social, and economic concerns to pursue a higher standard of living. People in prosperous community/society/country, for instance, need nourished food, energy, drinkable water, and clean air to live. A healthy environment is needed to do this. What distinguishes "sustainable development" from "sustainability" is that sustainability is a long-term goal (i.e., a more sustainable planet). In contrast, SD refers to the various processes and pathways that ultimately led to it (e.g., sustainable food production, green manufacturing, responsible consumption, good governance, research in green technology, its transfer, quality education and training, and so on). Thus far, education for SD was infused into several national and international frameworks and treaties designed to address key aspects of SD[11]–[15].

Nigeria is a country in West Africa with a population of 200 million and a GDP of $500 billion. Because of its heavy reliance on crude oil and natural gas exports, Nigeria's economy is considered mono-economic[21], [22]. As a result, Nigeria must diversify and transform its economy to achieve sustainability and the 17 SDGs through adequate infrastructure projects (dams, roads, airports, rail, solar-power, ICT, etc.)[23]. Infrastructural deficiencies impede socio-economic, techno-economic and socio-environmental development[24], [25]. Despite policies directed at the SDGs and several statutory guidelines and codes, SD standards and enforcement are weak in the built and economic environment[26], [27]. Sustainable funding for SD and green projects is low due to budget policies which exacerbates the efforts to address the social, environmental, and economic challenges in the built environment.

A review of extensive literature by Ogbonna[2] led to the identification and assessment of 14 factors that are determinant and requisite for SD and green economic development in Nigeria. The result highlights the impacts of development goals, frameworks, policies, and stakeholders' perceptions. Such factors are deemed critical to the economic success of the economy and SD in Nigeria. These were further supported by studies[8]–[13] emphasizing green/sustainable production and responsible consumption as factors critical to SD. In addition, past and recent research works[9], [17], [28], [29] stressed the importance and roles of concerned stakeholders in green, social, economic and SD. Such roles involve buy-in of the SD concepts, adherence and promoting SD, research and development of SD and green products.

Summarily put, this study identified and structures 11 SFs as follows: Socio-political and socio-economic stability (SF 1); Effective SD Policy and Legal Framework (SF 2); Political/Leadership will and SD enforcement (SF 3); Sustainable funding for SD projects and Infrastructure (SF 4); Concerned Stakeholder's buy-in and adherence to SD policies and framework (SF 5); Research and Development for innovative green technologies (SF 6); Job and wealth creation due from adopting, assimilating and utilizing green technologies (SF 7); Sustainable and green efficiency rating of industries and products (SF 8); Sustainable production, sustainable supply networks and responsible consumption (SF 9); Green tax and green rewards to encourage SD (SF 10); Public awareness of the impacts of SD targets and SD Goals (SF 11). These SFs were subjected to a tentative hypothetical assumption that they may be or may not be significant to the overall SD linked to Nigeria's economic growth and development.

4. Analyses and Results

4.1Research Responses and Respondents Demography

The research data was quantitative data based on a questionnaire survey. Table 1 shows the results from the pilot study conducted across 30 participants. It also shows that five SFs were strongly agreed, and six 6 SFs were agreed to affect the success and failure of SD linked to Nigeria's economic growth and development. The overall survey involved the distribution of 900questionnaires, of which: 708questionaire were responsive and retrieved, but only 650questionnaires were completely filled and used for analyses; the 58questionnaires that were uncompleted were left out of the analyses;192questionnaires were non-responsive. This shows that there 72.2% of responsive and analyzed questionnaires.

The respondent demography shows that the private sector constitutes 39.5% while the public sector constitutes 60.5% of responses. Respondents with undergraduate and below qualifications constitute 54.2% whereas those with post-graduate qualifications constitute 45.8%; respondents with below 10years of experience were 48%; with over 10years of experience were 52%; all of which males were 58.9% and females were 41.1%. The average reliability test for all the scores on the 5-point scale is 0.88, which is deemed very good.

4.2Evaluation Results of the SFs for SD in Nigeria

The overall evaluations of the eleven SFs for SD in Nigeria are in table 1 below based on the 650questionnaires analyzed using the mean item/average score (MIS) model, relative/ranking importance index (RII) model and also the T-test analyses.

Table 1. Evaluations of Success Factors for SD in Nigeria

S/N	Success Factors for SD in Nigeria	Pilot Study (30 respondents)		Research Survey Results (650 respondents)			
		MIS	*Remark*	*MIS*	*Remark*	*RII*	*Ranking*
SF1	Socio-political and socio-economic stability	4.57	Strongly Agree	4.31	Strong Impact	0.86	3
SF2	Effective SD Policy and Legal Framework	4.67	Strongly Agree	4.40	Strong Impact	0.88	2
SF3	Political/Leadership will and SD enforcement	4.40	Agree	3.53	Strong Impact	0.71	5
SF4	Sustainable funding for SD projects and Infrastructural Development outlook	4.73	Strongly Agree	4.17	Strong Impact	0.83	4
SF5	Concerned Stakeholder's buy-in and adherence to SD policies and framework	4.77	Strongly Agree	4.59	Extreme Impact	0.92	1
SF6	Research and Development for innovative green technologies	4.43	Agree	3.20	Moderate Impact	0.64	6
SF7	Job and wealth creation due from adopting, assimilating and utilizing green technologies	4.80	Strongly Agree	2.78	Moderate Impact	0.56	8
SF8	Sustainable and Green efficiency rating of industries and products	4.10	Agree	2.75	Moderate Impact	0.55	9
SF9	Sustainable production, sustainable supply networks and responsible consumption	4.07	Agree	3.14	Moderate Impact	0.63	7
SF10	Green tax and rewards to encourage SD	3.83	Agree	2.28	Low Impact	0.46	10
SF11	Public awareness of the impacts of SD targets and SD Goals	3.77	Agree	2.14	Low Impact	0.43	11

Source: Author's illustrations of the statistical assessments of the SFs.

The assessments in table 1 above indicate that one SF was deemed to have an extreme impact on the SD, leading to economic growth and development in Nigeria. Four SFs have a strong impact; four SFs with a moderate impact; two SFs with a low impact on the SD in Nigeria. The top-ranked factors were Concerned Stakeholder's buy-in and adherence to SD policies and framework(1st); Effective SD Policy and Legal Framework(2nd); Socio-political and socio-economic stability(3rd); Sustainable funding for SD projects and Infrastructural Development outlook(4th); Political/Leadership will and SD enforcement(5th). The middle-ranked SFs were Research and Development for innovative green technologies(6th); Sustainable production, sustainable supply networks and responsible consumption(7th); Job and wealth creation due from adopting, assimilating and utilizing green technologies(8th); Sustainable and green efficiency rating of industries and products(9th). The lowest-

IOP Conf. Series: Earth and Environmental Science 1044 (2022) 012012 doi:10.1088/1755-1315/1044/1/012012

ranked SF were green tax and rewards to encourage SD (10th); Public awareness of the impacts of SD targets, and SD Goals(11th). Table 2 below shows the testing of the research hypothesis using a T-test based on MIS evaluations of the SFs.

Table 2. T-test of Research Hypothesis

SFs	Mean	Standard Deviation	Standard Error	N	DF	Alpha (level of Significance)	P-value	Tcal	$Ttab_{0.05, 10}$	Significance
11nr. SFs	3.39	0.87	0.26	11	10	0.05	0.00	12.89	1.81	yes

Source: Author's illustrations of the statistical assessments of the SFs.

The hypothesis test indicates that with a 5% level of significance, P-value of 0.000, T-test calculated as 12.89, and T-test tabulated as 1.81, the 11 SFs identified and analyzed are significant to the SD that leads to economic growth and development in Nigeria.

4.3Implications of the evaluation Results of the SFs for SD in Nigeria

The implications of the above analyses on the eleven SFs affecting SD that leads to economic growth and development in Nigeria are that for developing countries like Nigeria to engage in and transform their economic activities per the UN SDGs, there is a need for workable and pragmatic SD policies and legal frameworks that are agreed upon and accepted by all stakeholders. In light of the present economic realities and the notion that SD policies and measures cannot be implemented effectively, lawmakers and leaders have put their hopes in sustaining most of the economy. However, there are serious issues with these targets and tactics. There will be no progress in eliminating poverty and correcting environmental degradation by adopting SD policies without considering and dealing with the many factors that influence these targets. Moreover, certain SD procedures may harm some industries, resulting in pushback from those sectors. Workable and practical methods to SD for Nigeria should fundamentally consider important stakeholders in main economic sectors, their operations, interest, and requirements to balance their influence on SD principles.

Funding for SD, particularly its infrastructure, requires long-term financing sources that can be relied upon. In order to continue the financing for initiatives that address the built environment's social, environmental and economic elements, a smart policy will allow a collaboration between public and private organizations to implement such projects. Nigeria has also created a green bond (GB). This environmentally friendly bond issued by Nigeria must comply with the existing international green bond criteria. The success of Nigeria's GB will establish a precedent for future SD projects. Leadership and political tenacity are required for implementing, enforcing, and maintaining the SD policy.

Despite the paradox in SD policy and enforcement, countries like Nigeria are still striving to industrialize to meet their economic demands with SD transition. As such, for the SD policy and legal framework to be fully embraced, there must be awareness within the majority of the population on the SD issues, impacts and long-term objectives. A new paradigm shift in consumer awareness of green labelled products/items will be ushered in due to this and pave the way for responsible consumption. Manufacturers and industries will be encouraged to use green and sustainable practices in their production and supply chains due to a shift in consumer behaviour leading to green and sustainable production. The resultant effects will enable green tax and rewards to encourage SD among organizations, producers, marketers/suppliers and consumers.

Through academic and industrial cooperation, SD efforts will enable research & development (R&D) of innovative smart and green technologies that will benefit Nigeria's economic growth. These technologies, if properly utilized, will aid in Jobs and wealth creation due to adopting, assimilating and utilizing such technologies.

Altogether, the suitability, consideration and applicability of such outlined success factors for SD in Nigeria will have a tremendous impact on Nigeria's success rate in achieving its 17 UN SDGs, its ranking in the Global Green Economy Index (GGEI) and the International green economy association (IGEA).

5. Conclusions

This research identified eleven SFs for SD in Nigeria's economy based on the stakeholders' perceptions. Mono quantitative method enabled the analyses of 650questionnaires. The result shows that these SFs are significant to the SD in Nigeria. It also shows the need for stakeholders' buy-in and cooperation for the SD policies, frameworks and efforts to be efficient in the following SD stages:

(1) Policy formulation: balancing all stakeholders' interests, needs, and influences; broad public awareness and cooperation with SD policies.Implementation: adopting green/sustainable production; sustainable supply networks; responsible consumption; public-private financing for SD projects; R&D for innovative, smart and green technologies;

(2) Enforcement: organizations adhere to green and SD models; a new paradigm shift where consumers' preferences go to green labels and green-rated products

(3) Adaptability: significant employment rate and wealth creation due to embracing, integrating, and employing green technology; green taxation and incentives for producers and buyers.

Summarily put, SD policies implementation, enforcement and adaptability may take years or a few decades, but once entrenched into overall economic sectors, organizational operations and corporate social responsibilities and the majority of the population's respective activities, the social, economic and environmental sustainability positive impacts will be profound.

References

[1] UNWCED, "Our Common Future: A Report by the UN World Commission on Environment and Development. The Brundtland Commission Report," 1987. [Online]. Available: http://www.un-documents.net/wced-ocf.htm.

[2] J. Ogbonna, "An Assessment of Determinant Factors for a Sustainable Green Economy in Nigeria," in IOP Conference Series: Earth and Environmental Science, 2021, vol. 793, no. 1, p. 12031.

[3] L. Kahle and E. Gurel-Atay, Communicating sustainability for the green economy. 2013.

[4] UNEP - UN Environment Programme, "What is an 'Inclusive Green Economy'?" https://www.unep.org/explore-topics/green-economy/why-does-green-economy-matter/what-inclusive-green-economy (accessed Feb. 05, 2021).

[5] Green Economy Coalition, "Green Economy: A Submission to the UN Conference on Sustainable Development zero draft text.," 2011.

[6] Global Network for Advanced Management, "Green Economy Management." https://globalnetwork.io/course/green-economy-management (accessed May 25, 2022).

[7] UK Government, "Enabling the transition to a green economy.," 2011.

[8] UNEP - UN Environment Programme, "Green Economy." https://www.unenvironment.org/regions/asia-and-pacific/regional-initiatives/supporting-resource-efficiency/green-economy (accessed Feb. 05, 2021).

[9] O. O. Adejumo, "Growth Limits: A Conceptual Analysis for Sustainable Development in Nigeria," SAGE Open, vol. 10, no. 2, p. 2158244020918277, 2020.

[10] A. Khatri, D. Garg, and G. S. Dangayach, "Critical success factors of sustainable manufacturing and procurement: an empirical study," International Journal of Social Ecology and Sustainable Development (IJSESD), vol. 10, no. 3, pp. 17–27, 2019.

[11] J. F. De Medeiros, J. L. D. Ribeiro, and M. N. Cortimiglia, "Success factors for environmentally sustainable product innovation: a systematic literature review," Journal of Cleaner Production, vol. 65, pp. 76–86, 2014.

[12] A. Zutshi and A. S. Sohal, "Adoption and maintenance of environmental management systems: critical success factors," Management of Environmental Quality: An International Journal, 2004.

[13] R. D. Raut, B. Narkhede, and B. B. Gardas, "To identify the critical success factors of sustainable supply chain management practices in the context of oil and gas industries: ISM approach," Renewable and Sustainable Energy Reviews, vol. 68, pp. 33–47, 2017.

[14] M. Hilton, Design for sustainable development. Success factors. European Foundation for the Improvement of Living and Working Conditions, 2001.

[15] UNESCO, "Sustainable Development," United Nations Education and Scientific Organization. https://en.unesco.org/themes/education-sustainable-development/what-is-esd/sd (accessed Jan. 16, 2022).

[16] UNEP - UN Environment Programme, "Green economy." https://www.unenvironment.org/explore-topics/green-economy (accessed Feb. 05, 2021).

[17] United Nations, "The sustainable development goals report," 2016. doi: 10.29171/azu_acku_pamphlet_k3240_s878_2016.

[18] D. R. Cooper and P. S. Schindler, Business Research Methods., 12th ed. The McGraw-Hill Companies, Inc., 2014.

[19] M. Saunders, P. Lewis, and A. Thornhill, Research methods for business students., 7th ed. Pearson Education Limited, 2016.

[20] R. V. Krejcie and D. W. Morgan, "Determining Sample Size for Research Activities.," Educational and Psychological Measurement., vol. 30, pp. 607-610., 1970.

[21] Z. A. Elum and A. S. and Momodu, "Climate change mitigation and renewable energy for sustainable development in Nigeria: A discourse approach," Renewable and Sustainable Energy Reviews, vol. 76, pp. 72–80, 2017.

[22] B. Lin and I. Ankrah, "On Nigeria's renewable energy program: Examining the effectiveness, substitution potential, and the impact on national output," Energy, vol. 167, pp. 1181–1193, 2019, doi: 10.1016/j.energy.2018.11.031.

[23] S. G. Dalibi et al., "Socio-Economic Performances of Mega Construction Projects (MCPs) in the Light of Sustainable Development of Nigeria's Built Environment," IOP Conference Series: Earth and Environmental Science, vol. 495, no. 1, pp. 1–9, 2020, doi: 10.1088/1755-1315/495/1/012046.

[24] Salisu Gidado Dalibi, "Green buildings and environmental sustainability issues in the Nigeria's built environment," in Nigerian Institute of Quantity Surveyors (NIQS) National Training Workshop held at At Ibom Hall, IBB Avenue, Uyo, Akwa Ibom State, under the Theme: "The Challenges of Environmental Sustainability and the Imperatives of Efficient Cost management in Nigeria, 2014, [Online]. Available: https://niqs.org.ng/.

[25] S. G. Dalibi and B. S. Bello, "Socio-economic impacts of Chinese government financed infrastructural development projects in Nigeria," in Second International Conference on Economic and Business Management (FEBM 2017), 2017, pp. 579–588.

[26] S. G. Dalibi, J. C. Feng, L. Shuangqin, A. Sadiq, B. S. Bello, and I. I. Danja, "Hindrances to Green Building Developments in Nigeria's Built Environment: 'the Project Professionals'' Perspectives",'" 2017, doi: 10.1088/1755-1315/63/1/012033.

[27] E. Barbier, "The policy challenges for green economy and sustainable economic development," in Natural resources forum, 2011, vol. 35, no. 3, pp. 233–245.

[28] M. Dittrich, S. Giljum, S. Lutter, and C. Polzin, "Green economies around the world," Implications of resource use for development and the environment, 2012.

[29] U. Inuwa Labaran et al., "Sustaining the Sustainable Development in Nigeria's Built Environment: Determinant Factors," IOP Conference Series: Earth and Envir0nmental Science, vol. 793, no. 1, Jun. 2021, doi: 10.1088/1755-1315/793/1/012029.

Examining the Relationship between Sustainable Finance and Green Economic Progress in Ghana: The Stakeholders' Perspectives

Eugene Misa Darko[1,*, a, b]

[1] School of Economics and Management Southeast University, 211189 Nanjing, China.

Corresponding Author: [*, a] 233179931@seu.edu.cn; [b]nanayawdarkoea@yahoo.com

Abstract. Every economic development and progress entails wealth creation for the benefit of citizens sustainably. Sustainable development requires public and private investment in green assets and infrastructure that decreases carbon emissions, increases energy and resource efficiency, and preserves biodiversity and ecosystem services for humanity. Ghana is striving to transform its economic activities according to GE standards, although it faces challenges in meeting these standards. Moreover, adequate funding is limited or unavailable, especially for developing countries. This study uses a questionnaire survey to examine the relationship between SF and GE factors in Ghana. SF incorporates social and environmental issues into financial decision-making and performance. The result shows that the SFFs and the GEFs are significantly related. Ghana needs to strengthen its SF and GE policies from top to bottom to achieve remarkable green economic transformation and smooth transition.

1. Introduction

Economic development and progress entail wealth creation leading to the economic welfare of citizens. Various stakeholders associate economic progress with job creation, industrialization, or infrastructure development. By contrast, economic progress includes all investment in major sectors of an economy leading to prosperity and quality of life. Economic development policies should include socio-economic, environmental, technological, and political factors to achieve significant economic progress. Several studies [1]–[4] have emphasized the unquestionable importance of sustainability, green economic development, and eco-friendliness in facilitating sustainable human interactions by creating a harmonious environment that balances the ecosystems, mitigates climate change and ensures a sustainable and healthy built environment (BE). While addressing citizens' socio-economic, techno-economic, and socio-environmental needs. Altogether, these factors contributed to the formation of the UN General Assembly's seventeen Sustainable Development Goals (SDGs) in 2015.

Policy analysts and researchers advocated the green economy concept to build a sustainable economy where society, the economy, and the environment harmoniously coexist [5]. GE concept is a socially inclusive, resource-efficient, and low-carbon development concept fueled by public and private investment in economic activities, green assets, and infrastructure. GE's concept aims to reduce carbon emissions and pollution, improve energy and resource efficiency, and ultimately prevent the loss of biodiversity and ecosystem services for humanity [6].

The concept advocates measures to mitigate ecological, environmental, and resource risks and scarcities to achieve sustainable development without causing harm to the ecosystem or the environment [7], [8]. GE further aims to improve the quality of life for humankind while maintaining the planet's ecological boundaries [9]. It optimizes value and growth throughout the economy while advocating for optimum management of natural resources sustainably [10]. In summary, GE encapsulates the whole economy as eco-friendly, sustainable, and resilient in all sectors typically linked with environmental products and services, such as renewable energy and other clean technology [5]. However, for GE's concept, approaches, measures, and policies to be effective, there must be adequate funds to sustainably finance green projects such as infrastructure and industries for responsible production, consumption, recycling, and other green, efficient services.

Sustainable finance (SF) contributes to long-term economic development and value creation. This implies that economic and social functions and people's way of life are secured and enhanced over the long term while maintaining the right cultural and social environment. In the financial industry, sustainable development contributes to achieving sustainable goals by including social, economic, environmental, and political leadership or governance in the value appraisal and risk management processes[11]–[13]. Additionally, SF encompasses impact investing and socially responsible investing by incorporating social and environmental factors into financial decision-making and performance [14]. Due to its consideration for socio-economic stability and maximizing sustainable development, SF is expected to evolve and assist investors in identifying high-value investment opportunities[15]. SF is becoming a more integral part of investors' and banks' risk management practices, especially in the new era involving foreign direct investment (FDI) [16]. Globalization and the resultant effects necessitate new innovative ideas for domestic investment, FDI, and initiatives that boost economic growth, protect the environment and help citizens, villages, towns, cities, and nations to prosper [17], [18].

The primary function of the financial system is to direct funds to major production areas. SF examines the connection between fiscal expenditure, economy, and environmental issues. These connections imply interlinkages between finance and sustainability in the short and long term [19]. SF also highlights various countries' divergent approaches to adopting GE and the strategies required to sustainably initiate, execute, and uphold GE in different economic sectors. To this end, Ghana is not an exception. As a developing country in Sub-Saharan Africa (SSA), Ghana strives to transform its economic activities according to GE standards. This transformation requires workable, pragmatic approaches within the financial system that can sustainably finance aspects of GE suitable to all the stakeholders. Although several studies [5], [20]–[24] have raised concerns over sources of sustainable funding for GE activities amid global economic unrest and financial crisis, politicians are still committed to implementing GE sustainable policies. However, Countries will struggle to achieve GE's goals of reversing environmental degradation and reducing poverty if political, economic, social, and cultural constraints are not fully considered and addressed through SF initiatives and measures.

Accordingly, this research examined the relationship between sustainable finance and the green economic progress of Ghana by identifying, assessing, and determining the correlation between the SF factors and the GE factors based on the perception of financial stakeholders and GE policy formulators [25]-[26].

2. Literature Review

Ghana is among the rapidly developing countries in SSA. It has a population of about a 33million people and a GDP of approximately US$73billion.The country's foreign reserve is about US$11billion. It has a human development index of 0.611 and a literacy rate of 80% (2018 estimate). The unemployment rate is 4.53% of the labor force. In terms of access to health and utilities, more than 40% of the total population have health insurance, and 87% have access to safe drinking water [27]. These social and economic indicators show the improvement of Ghana's efforts toward achieving the SDGs[28]. However, Ogbonna [5] opined that transitioning to GE may be less complicated in some developed countries, with fewer hurdles than in developing countries such as Ghana. From 2014 to 2018, the Global Green Economy Index (GGEI) placed Ghana in 15th – 20th position in Africa and between 46th -56th

globally out of 130 nations, while the 2019-2020 reports put Ghana in 11th in Sub-Saharan Africa [29]. Despite the improved ranking, Ghana's GE transition is relatively slow. The slow transition is attributed to poor implementation of factors accelerating GE and the SF.

2.1. Sustainable Finance and Development in Ghana

Globalization has intensified business competition and pollution through activities such as industrialization. Sustainability is now deemed crucial for the earth's continued existence. Sustainability entails sustainable production and responsible consumption while using reusable, recyclable materials with minimal energy input to produce high-quality products with minor environmental damage. A sustainable manufacturing system creates a balance between the economy and the environment. Environmental responsibilities are mandated by law and may offer green companies a competitive advantage [30].

The environment, society, culture, and economics are the four dimensions of SD. These dimensions are inextricably tied and therefore cannot be separated. Sustaining these dimensions requires sustainable financing. Ghana is presently diversifying and transforming its economy to achieve SDGs number 17. Such efforts aim to promote socio-economic, techno-economic, and socio-environmental development [28]. However, sustainable financing for SD and green projects is low to moderate [31]. Previous research works[28], [31]–[36] have attributed the performance of SF in a green economy to the performance of the following factors; green monetary policy, robust banking & financial system, green bond, financial aids, and FDI for green initiatives. Other SF factors include public-private funding, green fin-tech, green loans and mortgage, green procurement, green financial incentives, green tariffs, and taxonomy. Accordingly, this study employed the following sustainable finance factors (SFF) based on the literature. Green Monetary Policy (SFF1), Robust Banking & Financial System (SFF2), Green bond (SFF3), Financial Aids and FDI for Green initiatives (SFF4), Public-Private Funding (SFF5), Green Fin-Tech (SFF6), Green Loans and Mortgage (SFF7), Green Procurement (SFF8), Green Tariffs and Taxonomy (SFF9), Green Financial Incentives (SFF10).

2.2. Green Economic Transformation and Transition Factors in Ghana

Past and recent studies [20]–[23] indicated that building a GE requires cooperation among all stakeholders and the political leadership to enact GE change policies and sustainable economic development initiatives. Ghanaian policymakers need a better understanding of their ecological scarcity, which necessitates a more thorough economic and scientific examination of lost benefits and policy ramifications. Taxes on financial and forex activities, international lending facilities, and other creative funding options should be explored [37].

Altenburg and Rodrik [20], Barbier [37], and Krugman[38] asserted that international policymakers recognize eco-friendly growth, energy efficiency, and renewable energy as complementarity to the success of a GE. Furthermore, green and climate policy economists agree on the specific fundamental of GE. Several studies [2] [5], [7]–[10], [20]–[24] have identified, examined, and discussed issues related to GE, green growth, green production, and consumption. Their findings show that the green economic ideology is based on the scientific concept of eco-centrism, which recognizes the biosphere as the origin of life and the need for the environment and civilization to coevolve. However, this requires transitioning to GE across all sectors with support from major stakeholders.

However, there is a paradox here; Ghana remains at the early stages of industrialization, and therefore GE transition and implementation may face opposition from other stakeholders. Thus a workable and pragmatic policy framework for GE in Ghana should embrace the needs, interests, and requirements of key players in major economic sectors by structurally balancing the impacts of the GE concepts [28], [31], [33]–[35]. A study by Ogbonna [5] led to the identification and assessment of 14 determinant factors for sustainable green economic development. The result highlights the impacts of development goals, frameworks, policies, and stakeholders' perceptions critical to the success of GE. Studies from [8]–[13] further support the finding, emphasizing that sustainable production and responsible consumption are crucial to GE. In addition, past and recent research works[3], [37] stressed the

IOP Conf. Series: Earth and Environmental Science 1044 (2022) 012013 doi:10.1088/1755-1315/1044/1/012013

importance and roles of concerned stakeholders in green, social, economic, and SD policy formulations. Such roles involve accepting the GE concepts, adherence and promotion of SD, research, and development of SD and green products.

In conclusion, this study identifies the following factors as determinants of Green Economy growth based on the literature; Social and economic stability (GEF1), Consistent green economic Policies and Frameworks (GEF2), leadership drive towards Green economic transformation (GEF3), Sustainable funding for green projects and services (GEF4), and Green infrastructure (GEF5). Other factors include Stakeholder's cooperation and public awareness of GE benefits (GEF6), Adopting and utilizing green technologies for industrial and residential purposes (GEF7), Green production, sustainable supply networks, and responsible consumption (GEF8), Green ratings, rewards, and tax for industries and green products to encourage SD (GEF9), Sustainable Business initiatives leading to Job and wealth creation (GEF10).

3. Research Methodology

The research adopted a sequential quantitative research process to collect and analyze the data. The sequential approach is shown in figure 1 below.

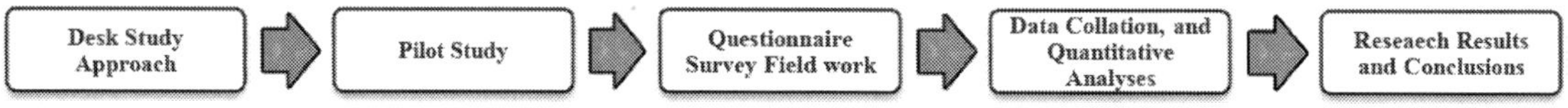

Figure 1. Research Methodological Framework
Source: Authors' Illustrations

This study conducted both desk and primary research. Desk research is mainly concerned with gathering secondary data from existing resources. Secondary data analyses entail reviewing previous research studies and reports, often disclosing much historical data or decision-making tendencies. Desk research helps identify existing problems at a low cost. It helps understand the fundamentals of an emerging situation or topic, for example, SF and GE in developing economies. The desk study provided preliminary insight into factors influencing GE and SF. These factors were used in the pilot study to ascertain their applicability in the Ghanaian context and fine-tune them for the questionnaire survey (primary research). Primary data helps provide a better understanding of the subject of SF and GE within the Ghanaian context. Moreover, the findings from the study are current, reliable, and appropriate for decision-making. Statistical models such as mean item score, relative importance index, and inferential statistics such as Pearson's correlation model and T-test were used to evaluate and determine the significance of such factors. The questionnaire survey was based on a random sampling technique [25], [26]; administered to the Ghanaian financial and GE stakeholders.

4. Data Presentation, Analyses, and Results

4.1. Analyses of Research Responses and Respondents' Demography

The study administered nine hundred and twenty-five (925) questionnaires, out of which seven hundred and two (702) respondents answered the questions correctly and returned the questionnaire representing 73.7% of the total sample. Two hundred fifty (250) respondents, representing 26.3%, did not return the questionnaire. Thus, the data analysis was conducted only for the seven hundred and two (702) respondents. The demography composition of the seven hundred and two respondents (702) includes 50.9% from the private sector, while the public sector constitutes 49.1%. 50.1% of the respondents have undergraduate or lower qualifications, whereas those with post-graduate qualifications constitute 49.1%. 51% of the respondents have less than 10 years of working experience, whereas 49% have more than 10years of experience. Also, 58.8% of the respondents were males, whereas 41.2% were females. The average reliability test for all the scores on the 5-point Likert scale is approximately 0.9, which is deemed very good.

4.2. Assessment Results of the SFFs and GEFs in Ghana

Table 1 presents the results of the pilot study conducted across 40 participants. The results show that respondents strongly agreed on four sustainable finance factors (SFFs), whereas they agreed on six of the SFFs factors. The six SFFs factors are critical to the success and failure of SD in Ghana. Consequently, a comprehensive survey was conducted to explore further the effects of these factors on Ghana's SF. A total of seven hundred and two (702) questionnaires were analyzed using the mean item/average score (MIS) model, relative/ranking importance index (RII) model, and the T-test analyses.

Table 1. Assessments of Sustainable Finance Factors for Green Economic Development in Ghana

	Pilot Study (40 respondents)		**Research Survey Results (702 respondents)**			
	MIS	***Remark***	***MIS***	***Remark***	***RII***	***Ranking***
Green Monetary Policy (SFF1)	4.50	Strongly Agree	2.60	Satisfactory	0.52	8
Robust Banking & Financial System (SFF2)	4.30	Agree	2.74	Satisfactory	0.55	6
Green Bond (SFF3)	4.20	Agree	4.53	Very Good	0.91	1
Financial Aids and FDI for Green initiatives (SFF4)	4.60	Strongly Agree	2.66	Satisfactory	0.53	7
Public-Private Funding (SFF5)	4.68	Strongly Agree	3.04	Satisfactory	0.61	4
Green Fin-Tech (SFF6)	4.70	Strongly Agree	3.32	Satisfactory	0.66	3
Green Loans and Mortgage (SFF7)	4.03	Agree	1.91	Poor	0.38	10
Green Procurement (SFF8)	3.83	Agree	3.96	Good	0.79	2
Green Tariffs and Taxonomy (SFF9)	2.55	Agree	3.01	Satisfactory	0.60	5
Green Financial Incentives (SFF10)	3.38	Agree	2.06	Poor	0.41	9

Source: Authors' Statistical Analyses

From the assessments in table 1, one SFF factor was deemed very good; six SFFs were satisfactory, whereas two SFFs performed poorly in promoting sustainable finance. The top-ranked factors were Green Bond (SFF3), Green Procurement (SFF8), and Green Fin-Tech (SFF6). The middle-ranked SFFs were Public-Private Funding (SFF5), Green Tariffs and Taxonomy (SFF9), and Robust Banking & Financial System (SFF2). The lowest-ranked factors were Financial Aids and FDI for Green initiatives (SFF4), Green Monetary Policy (SFF1), Green Financial Incentives (SFF10), and lastly, Green Loans & Mortgage (SFF7). Table 2 below presents the results of the GEF assessment in Ghana.

Table 2. Assessments of Green Economy Factors in the Economic Development of Ghana

Factors for GE in Ghana	**Pilot Study (40 respondents)**		**Research Survey Results (702 respondents)**			
	MIS	***Remark***	***MIS***	***Remark***	***RII***	***Ranking***
Social and economic stability (GEF1)	4.70	Strongly Agree	3.78	Good	0.76	1
Consistent green economic Policies and Framework (GEF2)	4.43	Agree	2.83	Satisfactory	0.57	7
Leadership will for Green economic transformation (GEF3)	4.58	Strongly Agree	3.04	Satisfactory	0.61	5
Sustainable funding for green projects and services (GEF4)	4.73	Strongly Agree	1.82	Poor	0.36	8
Green infrastructure (GEF5)	4.60	Strongly Agree	1.80	Poor	0.36	9
Stakeholder's cooperation and public awareness of GE benefits (GEF6)	3.67	Agree	3.35	Satisfactory	0.67	2
Adopting and utilizing green technologies for industrial and residential purposes (GEF7)	3.73	Agree	3.14	Satisfactory	0.63	4
Green production, sustainable supply networks and responsible consumption (GEF8)	3.58	Agree	1.78	Poor	0.36	10
Green ratings, rewards, and taxes for industries and green products to encourage SD (GEF9)	3.05	Agree	3.15	Satisfactory	0.63	3
Sustainable Business initiatives leading to Job and wealth creation (GEF10)	4.58	Strongly Agree	3.04	Satisfactory	0.61	6

Source: Authors' Statistical Analyses

The assessments from Table 2 indicate that one GEF showed good performance; six GEFs had satisfactory performance; while three GEFs demonstrated poor performance in promoting green transformation, green economic growth, and development in Ghana. The top-ranked factors were Social and economic stability (GEF1), Stakeholder's cooperation and public awareness of GE benefits (GEF6), Green ratings, rewards and tax for industries, and green products to encourage SD (GEF9). The middle-ranked SFFs were Adopting and utilizing green technologies for industrial and residential purposes (GEF7), Consistent green economic Policies and Frameworks (GEF2), and Sustainable Business initiatives leading to Job and wealth creation (GEF10). The lowest-ranked Factors were Sustainable funding for green projects and services (GEF4), Green infrastructure (GEF5), and lastly, Green production, sustainable supply networks & responsible consumption (GEF8). Table 3 below reports the test of significance for the factors.

Table 3. Test of Statistical Significance of the SFFs and GEFs

Factors	Mean	SD	SE	N	D F	Alpha	Pvalue	Tcal	$Ttab_{0.05,}$	Significance
10nr. SFFs	2.98	0.80	0.25	10	9	0.05	0.00	11.75	1.83	Yes
10nr. GEFs	2.77	0.72	0.23	10	9	0.05	0.00	12.22	1.83	Yes

Source: Authors' Statistical Analyses

Table 3 shows that all the diagnostic tests for SFFs and GEFs are statistically significant at the 1% significance level. For SFFs, the t-test calculated has a statistic value of 11.75, and the t-test tabulated at 1.83 at the 1% significance level. The t-test calculated a value of 12.22, and the t-test tabulated value of 1.83 is significant at 1% for the GE factors. The hypothesis test of the favorable effect of SFFs factors is statistically significant at the 1% level, suggesting all the factors of SFF are instrumental in promoting sustainable financing in Ghana. The result for the green economy factors is statistically significant at the 1% level (P-value, 0.00), implying favorable effects of GE factors in promoting a sustainable GE environment in Ghana. Table 4 presents the correlation analyses of the SFF and the GEF factors based on analyses from tables 1-3.

4.3. Assessing the Relationship between Sustainable finance and Green Economy factors in Ghana

Table 4 presents the statistical correlation results based on the assessed SFF and GEF factors in tables 1 and 2.

Table 4. Examining the Correlations between the SFFs and the GEFs

Correlation	*SFF1*	*SFF2*	*SFF3*	*SFF4*	*SFF5*	*SFF6*	*SFF7*	*SFF8*	*SFF9*	*SFF10*
GEF1	-0.34	-0.78	0.69	-0.82	0.54	0.48	-0.86	0.99	-0.10	-0.78
GEF2	0.38	0.94	-0.33	-0.01	-0.12	-0.82	0.87	-0.43	0.67	0.63
GEF3	0.17	0.45	0.40	-0.64	-0.17	-0.77	0.32	0.22	0.47	0.13
GEF4	0.61	0.93	-0.67	0.54	-0.35	-0.77	0.93	-0.85	0.21	0.93
GEF5	0.53	0.95	-0.71	0.58	-0.31	-0.70	0.96	-0.87	0.30	0.90
GEF6	-0.64	-0.80	0.21	-0.04	0.40	1.00	-0.73	0.45	-0.18	-0.76
GEF7	-0.02	-0.69	-0.18	0.07	0.69	0.79	-0.67	0.33	-0.46	-0.32
GEF8	0.84	0.81	-0.61	0.40	-0.24	-0.84	0.76	-0.69	0.00	0.98
GEF9	-0.56	-0.45	-0.04	-0.10	0.67	0.86	-0.40	0.35	0.27	-0.60
GEF10	-0.22	0.37	0.00	-0.47	0.49	-0.10	0.30	0.22	0.92	-0.13

Source: Authors' Statistical Analyses

Table 4 measures the correlation between the sustainable finance factors and the green economic factors according to the following; very strong (0.9 to 1.0), strong (0.7 to 0.9), moderate (0.5 to 0.7), low (0.3 to 0.5), negligible (0.0 to 0.3). The positive sign indicates a positive correlation, whereas the negative sign indicates a negative relationship. The results show a strong to a low but significant positive relationship between sustainable financing and a green economy, suggesting that Ghana can successfully transition towards a green economy with sustainable investment. For example, the effectiveness of green monetary policy in Ghana has a robust positive effect on green production, sustainable supply networks, and responsible consumption. It can be achieved via green incentives, rewards, and taxes. The study also found sustainable funding in green infrastructure and other green projects and services to stimulate green economic growth. The favorable effect of sustainable financing is directly linked to the favorable monetary policies, the robustness of Ghana's banking & financial system, the performance of its green bond, quality green procurement standards, the green fin-tech industry, green taxes, and other financial incentives. The study found a moderate effect of international aid and FDI suggesting that Ghana has not attracted large-scale FDI into green projects. These sustainable finance alternatives significantly affect sustainable business initiatives leading to job and wealth creation, green production, sustainable supply networks, and responsible consumption. On the other hand, the study found almost insignificant effects of sustainable financing on leadership and stakeholder interests suggesting the lack of political will towards initiating sustainable financial programs and financial accountability. It is important to note that sustainable financing and green transformation thrive on good leadership.

5. Conclusion and Policy Implications

The emergence of GE signals a transition to sustainably greening production and consumption from the traditional economy. However, building a green economy requires sustainable financing; this research examined the relationship between sustainable finance and the green economic progress of Ghana. The study examined 10 SFFs and GE factors in line with the literature. The study administered nine hundred and twenty-five (925) questionnaires; seven hundred and two (702) responded and were included in the analysis. The reliability test for the obtained data was deemed very good (0.9). The estimation results show that the 10 SFF factors have significant positive effects on the corresponding 10 GE factors implying that sustainable financing positively influences Ghana's green economic transformation, growth, and development. Overall, the correlation analyses show positive associations between sustainable finance and green economic factors. The study highlights the importance of quality leadership and stakeholder involvement in the green movement and the low carbon economy.

Moreover, the effectiveness of this transition requires the cooperation of stakeholders from top to bottom, including the government, environmental regulatory agencies, industrialists, manufacturers, professionals, academic researchers, environmentalist pressure groups, and civil society organizations with vested interests in sustainable development. GE programs may interfere with Ghana's massive investment in infrastructure and one district one factory industrialization programs. Thus, greening policy formulations should address stakeholder concerns, including trade-offs between conventional and green economies. SF examines socially responsible investments by including environmental and social factors in financial choices and decisions. However, due to financial constraints in several developing economies, policymakers are advised to design attractive investment policies, including green taxes and tariffs, to attract domestic and foreign investors. Eco-friendly air quality and forests health standards should be initiated to sustain economic growth. Lastly, the Ghanaian government should integrate all sectors of the economy with green policies to provide a habitable environment that balances social, economic, and environmental sustainability. The integration will reduce waste in the system and increase efficiency. Moreover, an enabling business environment with adequate or sustainable finance approaches facilitates green developments.

References

[1] UNWCED, "Our Common Future: A Report of the UN World Commission on Environment and Development. The Brundtland Commission Report," 1987. [Online]. Available: http://www.un-documents.net/wced-ocf.htm.
[2] M. F. Cordova and A. Celone, "SDGs and Innovation in the Business Context Literature Review,"

Sustainability, vol. 11, no. 24, p. 7043, 2019.
[3] United Nations, "The sustainable development goals report," 2016. doi: 10.29171/azu_acku_pamphlet_k3240_s878_2016.
[4] S. G. Dalibi *et al.*, "Socio-Economic Performances of Mega Construction Projects (MCPs) in the Light of Sustainable Development of Nigeria's Built Environment," *IOP Conference Series: Earth and Environmental Science*, vol. 495, no. 1, 2020, doi: 10.1088/1755-1315/495/1/012046.
[5] J. Ogbonna, "An Assessment of Determinant Factors for a Sustainable Green Economy in Nigeria," in *IOP Conference Series: Earth and Environmental Science*, 2021, vol. 793, no. 1, p. 12031.
[6] UNEP - UN Environment Programme, "Green Economy." https://www.unenvironment.org/regions/asia-and-pacific/regional-initiatives/supporting-resource-efficiency/green-economy (accessed Feb. 05, 2021).
[7] L. Kahle and E. Gurel-Atay, *Communicating sustainability for the green economy*. 2013.
[8] UNEP - UN Environment Programme, "What is an 'Inclusive Green Economy'?" https://www.unep.org/explore-topics/green-economy/why-does-green-economy-matter/what-inclusive-green-economy (accessed Feb. 05, 2021).
[9] Green Economy Coalition, "Green Economy: A Submission to the UN Conference on Sustainable Development zero draft text.," 2011.
[10] UK Government, "Enabling the transition to a green economy.," 2011.
[11] R. P. Gregory, J. G. Stead, and E. Stead, "The global pricing of environmental, social, and governance (ESG) criteria," *Journal of Sustainable Finance & Investment*, pp. 1–20, 2020, doi: 10.1080/20430795.2020.1731786.
[12] M. W. Sherwood and J. L. Pollard, "The risk-adjusted return potential of integrating ESG strategies into emerging market equities," *Journal of Sustainable Finance & Investment*, vol. 8, no. 1, pp. 26–44, 2017, doi: 10.1080/20430795.2017.1331118.
[13] K. Schumacher, H. Chenet, and U. Volz, "Sustainable finance in Japan," *Journal of Sustainable Finance & Investment*, vol. 10, no. 2, pp. 213–246, 2020, doi: 10.1080/20430795.2020.1735219.
[14] D. Schoenmaker and W. Schramade, *Principles of sustainable finance*. Oxford University Press, 2018.
[15] D. Schoenmaker, "A Framework for Sustainable Finance," *SSRN Electronic Journal*, 2018, doi: 10.2139/ssrn.3125351.
[16] G. Landi and M. Sciarelli, "Towards a more ethical market: the impact of ESG rating on corporate financial performance," *Social Responsibility Journal*, vol. 15, no. 1, pp. 11–27, 2019, doi: 10.1108/srj-11-2017-0254.
[17] C. A. Vargas, "Essays in sustainable finance," Jan. 2019, doi: 10.5167/UZH-175239.
[18] S. G. Dalibi and B. S. Bello, "Socio-economic impacts of Chinese government financed infrastructural development projects in Nigeria," in *Business and Management Research, volume 33 Second International Conference on Economic and Business Management (FEBM 2017)*, 2017, pp. 579–588.
[19] D. Schoenmaker, "Investing for the common good: A sustainable finance framework," *Brussels: Bruegel*, vol. 80, 2017.
[20] T. Altenburg and D. Rodrik, "Green industrial policy: Accelerating structural change towards wealthy green economies," *Green Industrial Policy*, 2017.
[21] M. Dittrich, S. Giljum, S. Lutter, and C. Polzin, "Green economies around the world," *Implications of resource use for development and the environment*, 2012.
[22] Z. C. Siamanta, "Building a green economy of low carbon: the Greek post-crisis experience of photovoltaics and financial'green grabbing'," *Journal of Political Ecology*, vol. 24, no. 1, pp. 258–276, 2017.
[23] C. Bann, P. B. Anand, U. Iftikar, T. Scott, G. Bouma, and S. Dutta, "Towards Green and Inclusive Prosperity: Building green economies that deliver on poverty reduction," 2015.
[24] U. Brand, "Green Economy; the Next Oxymoron? No Lessons Learned from Failures of

Implementing Sustainable Development," *GAIA - Ecological Perspectives for Science and Society*, , vol. 21, no. 1, pp. 28–32, 2012, doi: 10.14512/GAIA.21.1.9.

[25] D. R. Cooper and P. S. Schindler, *Business Research Methods.*, 12th ed. The McGraw-Hill Companies, Inc., 2014.

[26] M. Saunders, P. Lewis, and A. Thornhill, *Research methods for business students.*, 7th ed. Pearson Education Limited, 2016.

[27] Trading Economics, "Ghana Economic Data." https://tradingeconomics.com/search.aspx?q=Ghana (accessed Feb. 19, 2022).

[28] E. B. Ali, V. P. Anufriev, and B. Amfo, "Green economy implementation in Ghana as a road map for a sustainable development drive: A review," *Scientific African*, vol. 12, p. e00756, 2021.

[29] Dual Citizen LLC, *2016 Global Green Economy Index*. Dual Citizen LLC, 2016.

[30] A. Khatri, D. Garg, and G. S. Dangayach, "Critical success factors of sustainable manufacturing and procurement: an empirical study," *International Journal of Social Ecology and Sustainable Development (IJSESD)*, vol. 10, no. 3, pp. 17–27, 2019.

[31] A. Ahenkan, J. Osei, and E. H. Owusu, "Mainstreaming green economy: An Assessment of private sector Led Initiatives in climate change adaptation in Ghana," *Journal of Sustainable Development*, vol. 11, no. 2, pp. 77–87, 2018.

[32] United Nations., "Transforming Our World: The 2030 Agenda for Sustainable Development," *A New Era in Global Health*. 2018, doi: 10.1891/9780826190123.ap02.

[33] P. B. Cobbinah, R. Thwaites, and R. Black, "Ecotourism as a mechanism for achieving a green economy in developing countries: Experiences from Ghana," in *Tourism in the green economy*, Routledge, 2015, pp. 247–263.

[34] J. Amankwah-Amoah and D. Sarpong, "Historical pathways to a green economy: the evolution and scaling-up of solar PV in Ghana, 1980–2010," *Technological Forecasting and Social Change*, vol. 102, pp. 90–101, 2016.

[35] C. Breisinger, X. Diao, J. Thurlow, and R. M. Al Hassan, "Potential impacts of a green revolution in Africa—the case of Ghana," *Journal of international development*, vol. 23, no. 1, pp. 82–102, 2011.

[36] C. Kirkpatrick, D. Parker, and Y.-F. Zhang, "An empirical analysis of state and private-sector provision of water services in Africa," *The World Bank Economic Review*, vol. 20, no. 1, pp. 143–163, 2006.

[37] E. Barbier, "The policy challenges for green economy and sustainable economic development," in *Natural resources forum*, 2011, vol. 35, no. 3, pp. 233–245.

[38] P. Krugman, "Building a green economy," *New York Times*, vol. 5, 2010.

Research on Pedestrian Space Disorder in Old City Living Streets: A Case Study on the Anzipian Area of Harbin

Ye Yang[1,2,*,a], **Jia Chaozhi**[1,2,b], **Zhao Tong**[1,2,c] **and Li Zekai**[1,2,d]

[1]School of Architecture, Harbin Institute of Technology, China
[2]School of Architecture, Harbin Institute of Technology; Key Laboratory of Cold Region Urban and Rural Human Settlement Environment Science and Technology, Ministry of Industry and Information Technology, China

E-mail: [*,a]yeyang@hit.edu.cn;[b]20s134182@stu.hit.edu.cn;[c]1098593599@qq.com; [d]986383408@qq.com

Abstract. Livable quality promotion is an indispensable guideline of inventory renewal, and the old city's living street quality is the main focus of attention in this context. Based on the physical disorder theory, this paper takes the Anzipian area of Harbin as the research object, conducts virtual audits through street view images, and quantifies the degree of pedestrian space disorder. Then, this paper analyses the main problem of pedestrian space disorder in the old city living streets and points out optimal design strategies for these problems. Finally, the enlightenment of old city living street renewal research was summarized and the future direction of street physical quality research was prospected.

1. Introduction
Over the past few decades, the extensive urban development of Chinese cities has led to serious problems with urban space quality, especially concerning physical environment deterioration and the poor maintenance of public spaces. The strongly advocated vehicle priority strategy ignores residents' living demands in the street. Additionally, the squeezed and occupied pedestrian space worsens the walking experience for pedestrians. Existing old city streets, which cannot meet the high-quality needs of the residents' living standards, urgently need to be optimized.

2. Background

2.1. Neighborhood disorder
Neighborhood disorder has emerged as a particularly prominent term in a collection of urban-related research and is described as a generally perceived lack of order and control. It is difficult to find a standard definition for the extensively used concept of "neighborhood disorder" and distinction is often drawn between social disorder and physical disorder. Defined as "the overt signs of negligence or unchecked decay," physical disorder includes a range of substandard street space quality, such as illegally parked vehicles, the presence of graffiti, buildings in states of disrepair, and public property damage. In contrast, social disorder is defined as "unsettling or potentially threatening and unlawful public behaviors," which includes a range of behaviors that are uncivilized, socially criminal, or disturb social order [1-4]. Physical disorder theory is still in its infancy in China, but there has been growing interest in applying physical disorder in studies focusing on the street quality. Based on the physical

disorder theory, Chen and Long [5] and Chen et al. [6] measured the quality of urban street space in Beijing and Hefei city. Xu [7] focused on the disordered urban-scape in Chinese cities and explored the conformity mechanism of their solution. Yuan and Chen [8] focused on illegal vehicle parking and examined how parking areas resolve disordered parking. In the process of promoting urban renewal, there is still a lack of attention on old city street disorder in different urban areas and in finding methods to identify and quantify physical disorders in old city streets.

2.2. Virtual audit through street view images

Disorder is traditionally measured through in-person collection. However, street view images represent an alternative source of data on neighbourhood environments, providing an effective and reliable alternative to in-person street audits within various fields [9-11]. Street view images provide observation at the eye level, and they are more appropriate for assessing the items for pedestrian environment assessment [12]. In general, comparative studies conclude that street view images are a reliable and valid instrument for measuring disorder at the neighborhood level. Compared to in-person observations, researchers can conveniently visit neighborhood streets virtually online. Furthermore, due to street view images exhibiting a powerful zoom and high resolution, neighborhood characteristics can be captured in 360° images. Users can view these virtual spaces as if they were physically traveling there. To summarize, the present study contributes to virtual auditing explicitly through street view images.

3. Methods

First, existing theoretical research from both China and abroad is summarized. It is combined with the development characteristics of the old city streets to produce a quantitative items system. Then, a virtual audit is conducted by obtaining network data and street view images. Third, the results are quantified and analysed. Finally, strategies that correspond to key issues are developed.

3.1. Physical disorder items selected for the pedestrian space of living streets in old city

Even though there are no specific requirements in China for the environmental evaluation of old city street disorder, some researchers have argued the potential correspondence between urban street design guidelines and street disorder items, summarizing the significance of optimizing street physical quality. Based on existing research, Chen et al. [13] first constructed a quantitative index system for general street spatial disorder in Chinese cities, which included 19 spatial disorder items in five dimensions. As for living streets in the old city, the physical environment items reflecting the usual features of streets established in old residential districts are necessary to capture the general physical disorder of pedestrian spaces. Although many of the characteristics were thought to be closely associated with disorder perceptions, the characteristics selected for the present study are slightly more advantageous. Their perceptual implications are generally more straightforward and easier to interpret. The present study provides a brief analysis of the existing problems that are believed to contribute to physical disorder in old city living streets.

According to descriptions of disordered environments in relevant references [1,4,13], the characteristic elements of street pedestrian spaces in an old city, and the feasibility of virtually auditing street view images, physical disorder is divided into three dimensions: pedestrian facade, pedestrian surface, and pedestrian node. The following discussion focuses on these three dimensions with nine items, with examples shown in Table 1.

The level of the living street facade, which encompasses the building facade, ground store facade, and vegetation facade, is referred to as the pedestrian facade. Vegetation serves as a soft facade element, while the other two serve as hard facade elements.

Pedestrian surface provides essential conditions for walking. Unlawful parking, private possessions on pedestrian surfaces, obstacle structures on pedestrian surfaces, and the insufficient width of the pedestrian path are all pedestrian obstacles relevant to surface disorder.

Pedestrian node dominates a fragmented walking experience. The node continuity experience supported by indicated road signs reminding pedestrians and drivers to regulate their behavior and manage themselves safely is of considerable importance. This is because pedestrians are affected by the traffic and the building group's entrance. The proposed road sign shortage and weak transferability to assess pedestrian node disorder is directly linked to pedestrian space disorder of the living street.

Table 1. Pedestrian space disorder dimensions and items

Dimension	Item			
Pedestrian facade disorder	 Building facade decay	 Ground store facade decay	 Vegetation facade decay	
Pedestrian surface disorder	 Unlawful parking on pedestrian path	 Private goods on pedestrian path	 Obstacle structure on pedestrian path	 Insufficient width of pedestrian path
Pedestrian node disorder	 Road sign shortage	 Weak transferability		

3.2. Virtual audits

In this study, four auditors with professional architecture or urban planning backgrounds were selected and randomly assigned to audit work after receiving unified training. Each point of the street sample is audited using the disorder items, and each item's appearance in one direction is counted as 1 score, whereas it is counted as 2 if the item's appearance is in both directions.

4. Case study area and data

The Anzipian area of Harbin city, China was chosen as the case study, as shown in Figure 1. This area is composed of 26 streets with names beginning with *An-*, such as Anjing Street, Ansheng Street, Andao Street and so on. These urban branch living streets are dominated by rapid-expansion residential buildings that have been built since the 1970s, dividing the district into small blocks. Due to the age of construction and lack of maintenance, many physical environmental disorder phenomena, such as the unkempt facades of buildings and illegally occupied walking spaces, greatly inconvenience and negatively affect the residents daily.

Based on the OpenStreetMap road network data, we select sampling points with a distance of 25m on each street in Anzipian. Street view images obtained from Baidu Map were utilized, with each sampling point capturing street view images from two directions (0 and 180°) that are parallel to both sides of the street at a fixed height, as shown in Figure 2. Finally, 1118 sampling points were selected and a total of 2236 street view images (1024*512 pixels) were obtained, as shown in Figure 3. The street view images for the selected points were recorded between November and December 2019.

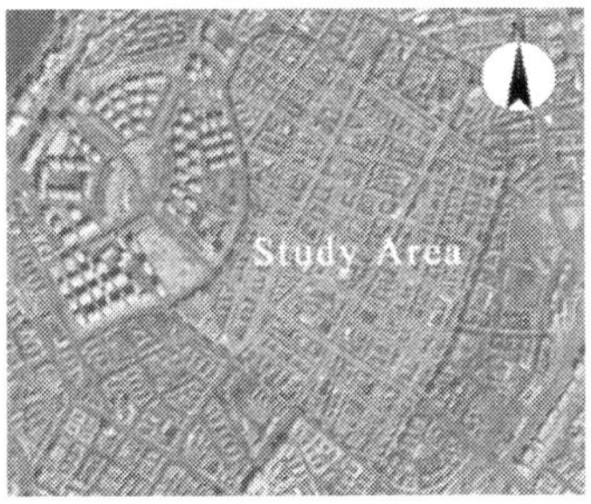

Figure 1. Case study area

Figure 2. Street view images in two directions

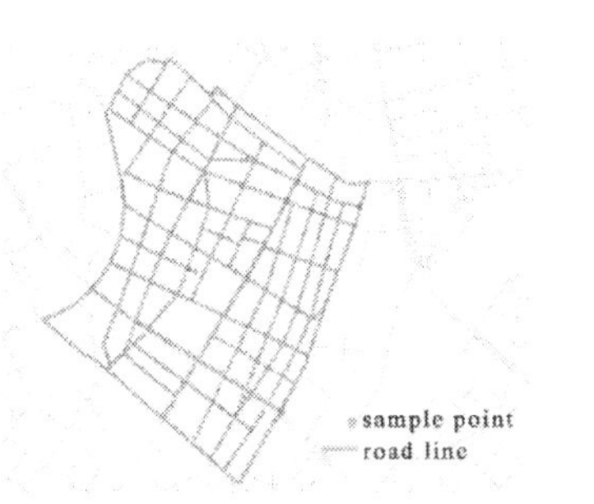

Figure 3. Selected sampling points

5. Results

5.1. Pedestrian space disorder distribution in Anzipian

The degree of pedestrian disorder in the research region is depicted. The color in the figure represents the range of disorder levels, where the green points represent a low degree, and the red points represent a high degree. Each point has a value ranging from 0 to 18 scores, with 0 indicating no disorder and a higher score indicating a higher degree of pedestrian space disorder. QGIS software was utilized for visual expression, and the scores were broken into five levels using the natural break approach. The following figures depict the average disorder score patterns of each level: Figure 4 shows the average score of each sample point, Figure 5 shows the average score of each street segment (divided by intersections), and Figure 6 shows the average score of single streets.

The sample sites were subsequently summed to acquire the resulting pattern on three levels, presenting the findings directly compared with each street. As can be observed, Anjing Street presents a more intense disorder phenomenon than the peripheral areas. This is in line with the fact that the local government has designated Anjing Street as a morning and evening market for locals, with disorder scores being the highest in the area centered on Anjing Street (Anjingerhutong is an extra section of Anjing Street). Many of the streets in this area are out of order. Additionally, disorderly features are especially common in the north section of Anding Street, except in the Anjing Street area. Anxiang Street and its surroundings have the least amount of disorder, and almost every street is green, as can be observed in the typical street view images in Figures 7, 8 and 9. Figure 10 shows scores for each street.

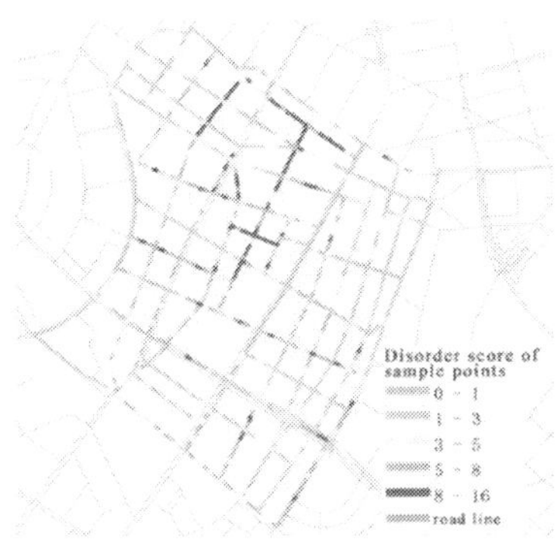

Figure 4. Disorder score of sample points

Figure 5. Disorder score of street segments

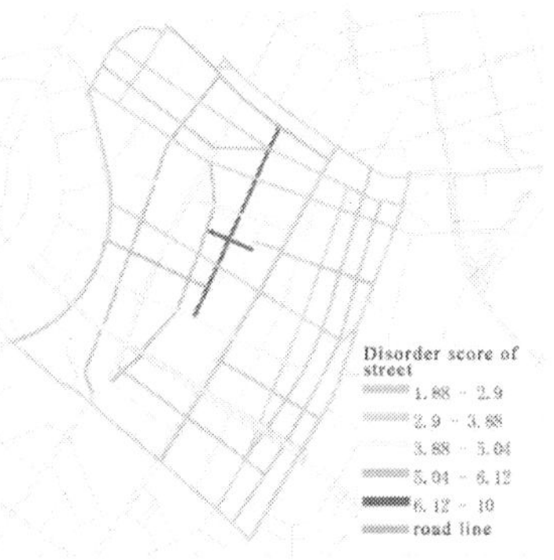

Figure 6. Disorder score of streets

Figure 7. Typical street view image of Anjing Street

Figure 8. Typical street view image of Anding Street

Figure 9. Typical street view image of Anxiang Street

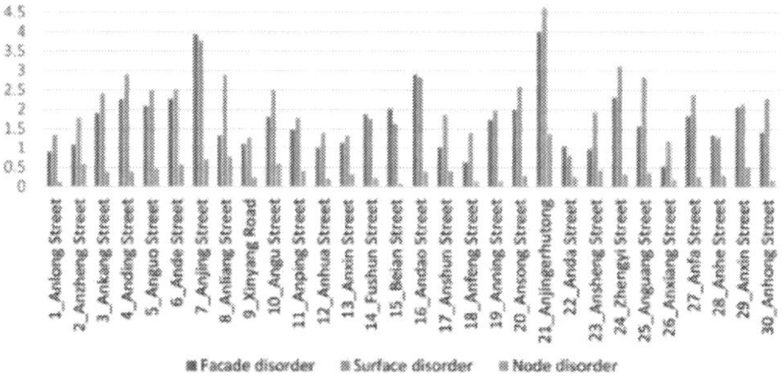

Figure 10. Disorder scores for each street

5.2. Pedestrian space disorder item distribution in Anzipian

Figure 11 shows the frequency of each disorder item. With a score of over 900, the disorder of the vegetation facade and unlawful parking were noticeable, whereas the other items stayed clustered below 600. Figure 12 shows the distribution of each disorder item.

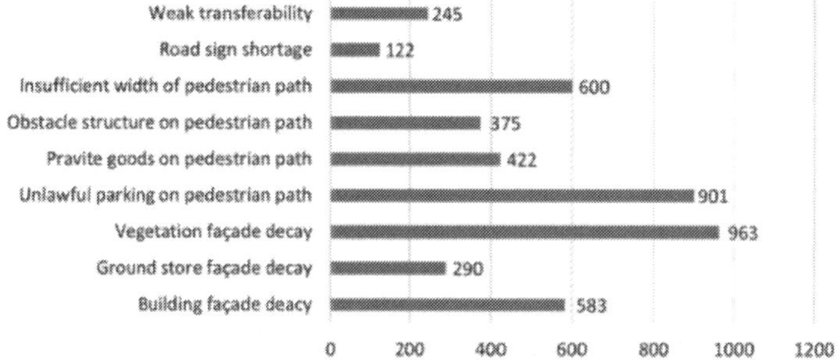

Figure 11. Disorder frequency of each item

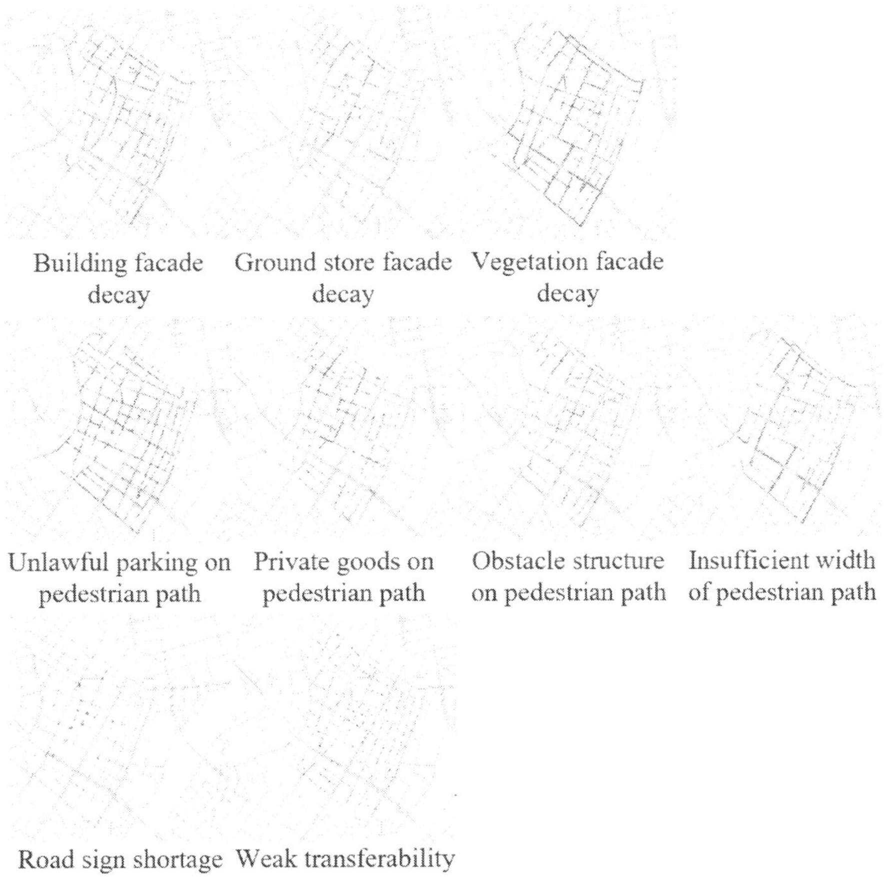

Figure 12. Distribution of each disorder item

5.3. Analysis of main disorder items

The disorder scores assess the overall environmental state of the pedestrian space. The individual disorder item generates the approach to increasing living quality. Optimizing each of these disorder items is an important method for meeting the needs of the residents and improving the quality of the city with the background of urbanization.

In the pedestrian facade dimension, the following was determined. 1) Due to the nature of the urbanization process, the old city building surfaces gradually deteriorate. While street facade renovations are always done on the main street, these dwelling streets that branch off from the main street suffer from a higher neglect rate. 2) With increasing motorized traffic and limiting construction conditions, some sidewalks have been transformed into vehicle lanes, primarily impacting sidewalk and tree lanes. These two disorders in old city living streets are common phenomena that residents sensitively perceive.

In the pedestrian surface dimension, the following was found. 1) Most residential quarters, which were built in the previous era but are still used today, did not anticipate car being as prevalent as they are now during the early stages of design, resulting in a shortage of parking spaces and a serious mismatch between the current situation and demand. This, in turn, further results in heavy street space occupation. 2) Private goods occupying pedestrian spaces to some extent represent an active street life that carries the daily activities of urban residents. However, respecting and balancing the spatial demands of all road users is essential for living street services. Private belongings and street vendor activities necessitate a flexible regulatory system to accommodate them in a more orderly manner. 3)

Over time, many of the buildings on street boundary that are used as dwellings have opened the ground floor doors and windows to conduct business. This transformed the building from being residential to commercial, promoting the installation of concrete steps and steel staircases on the ground level and causing the pedestrian space in front of the building to be occupied. 4) Mechanically widening the vehicle width during the rapid urbanization period leads to inadequate service level, insufficient breadth, and discontinuous pedestrian amenities.

In the pedestrian node dimension, even though these two items are important for the walking experience, they did not score well in this study. This is true not only for being dependent primarily on the number of intersections and entrances, but also regarding the size of assignment in the audit process and the number of times these phenomena appear in the street view images.

6. Design optimization strategy for pedestrian space disorder of living street in an old city

This research proposes the following strategies and guidance for inventory renewal concerning optimizing pedestrian space quality, highlighting living streets, and planning for local residents.

1) Ensuring the renewal value for residents and strengthening pedestrian facade repair. The process of upkeeping and updating the contents of living streets should focus on the progression of local features, humanity, and cultural values. The restoration and maintenance of living streets in old cities are inseparable from the in-depth research and analysis of old urban areas. On the one hand, it is suggested that conducting in-depth research and analysis will help with understanding the demands of the residents. On the other hand, the landing's design should emphasize the street's qualities in the living area while also ensuring that the facade is balanced. The rational replenishing of street trees, flower beds, and other greenery should also be a focus when updating the local features, which can help preserve and repair the present green facade, enhance plant color levels, and beautify the appearance of the street.

2) Strengthening pedestrian space management and improving street operating efficiency. Livable processes, experiences, and perception should be fully supported in optimizing inventory renewal. The sidewalk should be wide enough to accommodate residents. Under the limitations of the existing road width, combined with the present road space allocation, traffic efficiency should be maximized in areas where the sidewalk width does not match the standard. Bicycle and motor vehicle parking belts that considerably influence pedestrian spaces should be re-planned. Furthermore, vendors and private goods should be managed and supervised to ensure that pedestrian space is uncluttered. Private goods in the front area of the building should be clearly separated from the pedestrian space by implementing easily identifiable ground signs, street furniture, and other means. These actions would assist with clearly distinguishing pedestrian activity spaces, strengthening the management of the pavement, and ensuring an effective pedestrian width. Any unlawful steel stairs or concrete steps that protrude from the building's front area should be demolished, and the designated boundary should be strictly controlled.

3) Installing a signage system and improving the efficiency of the passage. Combing the ideas of efficient and sustainable transport, the design of the specified road area, routes, and rights can be conducted. With design methods such as adding signage systems, redefining the pedestrian area, and painting the road, the planning of passage crossing efficiency can be improved. Additionally, supplementary directing signs should be installed at nodes to remind people and vehicles.

In empirical research, it should not be assumed that a renewal strategy that is successful in one city will have the same outcome in another. While the common problems associated with deterioration and poor maintenance are similar, each old city area is different. A comprehensive and refined analysis of the neighborhood study object needs to be accompanied by refined assessments and conditions to make an informed decision.

7. Conclusion

This paper focuses on the pedestrian space disorder of living streets in the old city. The study was conducted by exploring the virtual audits through street view images and producing nine items to quantify the physical disorder. Through empirical research, it is found that a virtual audit is an efficient and valid method for characterizing the physical environment of living streets in old city area.

Furthermore, it is determined that the measurement of disorder can provide an important basis for refined management and inventory renewal in the near future.

The globally available and publicly accessible street-level images would greatly benefit old cities' urban environmental studies at the street level. Moreover, with the rapid developmental progress in artificial intelligence research, obtaining more information is becoming increasingly possible from street-level images, allowing different urban features along the streets to be accurately recognized. Furthermore, the characteristics chosen to interpret physical disorder will become more straightforward and easier to examine. The anticipated autonomous technologies will provide ample methods for refined planning and management in the future.

This research is still weak in the following aspects. Firstly, the street physical environment reflected by street view images has a lagging influence. As such, up-to-date street view imaging is needed. Secondly, some images may cause errors in the audit process due to unexpected shelter. Finally, the selected items can be further refined. Future studies should address these shortcomings to provide a more comprehensive understanding of pedestrian space disorder in old city living streets.

Acknowledgment

This paper was funded by the National Natural Science Foundation (Youth Fund) of China (Grant No.52008128).

References

[1] Ndjila S, Lovasi Gina S, Fry D and Friche A A 2019 Measuring Neighborhood Order and Disorder: a Rapid Literature Review *Current environmental health reports* **6** 316-326.

[2] Ross C E, Mirowsky J 1999 Disorder and decay: The concept and measurement of perceived neighbourhood disorder *Urban Affairs Review* **34(3)** 412-432

[3] Skogan W G 2012 Disorder and Crime *The Oxford handbook of crime prevention* Oxford University Press 175

[4] Marco M, Gracia E, Martín-Fernández M and López-Quílez, A 2017 Validation of a Google Street View-based neighbourhood disorder observational scale *Journal of urban health* **94(2)** 190-198

[5] Chen J and Long Y 2021 Element Identification, Measurement, Impact Evaluation and Spatial Intervention of Disordered Urban Public Space *Time+ Architecture* **01** 44-50 (in Chinese)

[6] Chen C, Chen J, Jia Z, Deng Z, Ma S, and Long Y 2020 Large-scale Measurement of the Quality of Urban Street Space Based on Physical Disorder Theory: A Case Study of Area within the Second Ring of Hefei City *South Architecture* **02** 10-18 (in Chinese)

[7] Xu G 2002 Disorder Urban-scape Connectors and Conformity Mechanism China *CNKI* (in Chinese)

[8] Yuan Q and Chen X 2020 Exploring Parking Governance and Public Service Improvement in the 15-Minute Community Living Circle *Communication & Shipping* **7(05)** 29-36 (in Chinese)

[9] Ye Y and Fei T 2020 Research on the Evaluation of Street Space Walkability Based on Street View Image *Contemporary Architecture* **11** 123-125 (in Chinese)

[10] Zhang L Y, Pei T, Chen Y J, Song C and Liu X 2019 A review of urban environmental assessment based on street view images *Journal of Geo-information Science* **21(1)** 46-58 (in Chinese)

[11] Thom S and Wim H 2021 Utilizing Geo-referenced Imagery for Systematic Social Observation of Neighborhood Disorder *Computers, Environment and Urban Systems* **90** 101691

[12] Mahdi A, Mehdi M, Muahammad Z S and Zohreh A 2018 Addressing issues in the use of Google tools for assessing pedestrian built environments *Journal of Transport Geography* **73** 185-198

[13] Chen J, Zhang Z and Long Y 2020 Strategies for Improving the Quality of Urban Street Space Oriented to Promoting Public Health: Perspective from Spatial Quality *City Planning Review* **09(44)** 35-47 (in Chinese)

AUTHOR INDEX